Translations

of

Mathematical Monographs

Volume 42

Direct and Inverse Imbedding Theorems. Applications to the Solution of Elliptic Equations by Variational Methods

by

L. D. Kudrjavcev

Lev Dmitrievich Kudri͡avt͡sev

American Mathematical Society
Providence, Rhode Island
1974

ПРЯМЫЕ И ОБРАТНЫЕ ТЕОРЕМЫ ВЛОЖЕНИЯ
ПРИЛОЖЕНИЯ К РЕШЕНИЮ ВАРИАЦИОННЫМ МЕТОДОМ
ЭЛЛИПТИЧЕСКИХ УРАВНЕНИЙ

Л. Д. КУДРЯВЦЕВ

Академия Наук
Союза Советских Социалистических Республик

ТРУДЫ
МАТЕМАТИЧЕСКОГО ИНСТИТУТА
именм В. А. СТЕКЛОВА
LV

Издательство Академии Наук СССР
Москва 1959

Translated from the Russian by S. Smith

AMS (MOS) subject classifications (1970).
Primary 35A05, 35A15, 35J20, 35J25, 35J70, 46E35;
Secondary 35B05, 35B30, 35B45, 35J67, 46A30, 35C15.

Library of Congress Cataloging in Publication Data

Kudrıavtsev, Lev Dmitrievich.
   Direct and inverse imbedding theorems.

   Translations of mathematical monographs, no. 42.
   Translation of Prıamye i obratnye teoremy vlozheniıa,
originally published as v. 55 of Trudy of Mathematiche-
skiı in-t of Akademiıa nauk SSSR.
   Bibliography: p.
   1. Differential equations, Elliptic--Numerical solutions. 2. Embedding theorems. I. Title. II. Series.
QA377.K813   515'.353   73-22139
ISBN 0-8218-1592-X

TABLE OF CONTENTS

INTRODUCTION

The solution of elliptic partial differential equations by the direct variational method has its origins in the works of Gauss, Thomas and Riemann on the investigation of Laplace's equation when considered as the Euler equation for the Dirichlet integral. But the basic principle underlying this investigation, which is known as Dirichlet's principle, was not rigorously proved until much later (1900–1901) in the works of Hilbert, which stimulated a whole series of investigations in this direction (by Courant, Levy, Fubini, Lebesgue, Zaremba and others). Important contributions to the development of these methods were made by Soviet mathematicians N. N. Bogoljubov, N. M. Krylov, M. A. Lavrent′ev, L. A. Ljusternik, S. L. Sobolev, V. I. Kondrašov and S. M. Nikol′skiĭ. A closely adjoining paper in a sense is that of M. V. Keldyš and M. A. Lavrent′ev [1] on the stability of the Dirichlet problem relative to a variation of the boundary of the domain. One should especially note the monograph of Sobolev [2] concerning, in particular, the investigation of Dirichlet's principle in n-dimensional space. It was shown in this work that every function having weak derivatives and a finite Dirichlet integral takes completely defined boundary values in the sense of convergence in the mean; the statement of the problem was thereby provided with a completely final form. In the same place are given necessary conditions which a boundary function must satisfy in order that the first boundary problem for a polyharmonic equation might be solved by a variational method in a bounded domain. Of significance in the development of this theory were the papers of Nikol′skiĭ [3–6], in which, in the case of domains with a smooth boundary, sufficient conditions coinciding "to within an arbitrary $\epsilon > 0$" with the necessary conditions established earlier by Sobolev were indicated in terms of boundary values for the applicability of the variational method to the above mentioned problem. A development of the variational method for more general elliptic equations, both linear and nonlinear, is given in the works of Kondrasov[7–10]. We also note the works of Aronszajn [46], Gagliardo [47], Slobodeckiĭ [48] and Babič and Slobodeckiĭ [11], in which necessary and sufficient conditions, which are imposed on a boundary function, are obtained for the possibility of applying the variational method to the solution of the first boundary problem for the harmonic and polyharmonic equations in the case of a finite domain with a sufficientlv smooth boundary; much more general problems are also considered.

The apparatus of the so-called *direct imbedding theorems*, which was created by Sobolev and his student Kondrašov, plays a large role in the application of the variational method to the problems of mathematical physics. For this reason the main part of the present monograph is devoted to the subsequent development of questions connected with *direct and inverse imbedding theorems*. The investigations of *inverse imbedding theorems* or, what is the same thing, *theorems on the extension of functions* and systems of functions conducted by us are a continuation of the corresponding investigations of Nikol'skiĭ [4].

On the basis of the results obtained on imbedding theorems we apply the variational method to the investigation of boundary problems for second order partial differential equations of elliptic type, both nondegenerate and degenerate, on the whole boundary of the domain or on part of it.

Here a new basis for *variational principles* is given which extends them to *arbitrary finite domains without any restrictions on the structure of the domains and their boundaries.* Existence and uniqueness theorems are proved for both the variational problem and the corresponding differential problem in the case of a selfadjoint elliptic equation of second order. The connection between the smoothness of the coefficients of an equation and the smoothness of a solution is studied.

In addition, proofs of the existence and uniqueness of the solution of boundary problems with sufficiently general boundary conditions are also given in the presence of a power degeneracy with exponent $\alpha > 0$.

In discussing these questions we have mainly followed a method whose foundations go back to Hilbert [12] and which has been developed and refined in the monograph of Sobolev [2] and in the works of Nikol'skiĭ [4–6] and Kondrašov [7–10].

The proof of the existence of a solution of the boundary problems for degenerate elliptic equations requires a study of the question of *weighted function extensions* from the boundary onto the whole domain. This is done in Chapter I of the present work. Here an essential role is played by both the methods of Nikol'skiĭ (for example, the method of approximating differential functions by entire functions of exponential type, the methods of function extensions and others) and the methods of Sobolev (for example, the method of function averagings), which naturally receive a further development and sharpening.

In this chapter we construct the *theory of best extensions* (in the sense of the growth of the derivatives under an approach to the boundary of the domain) of functions from the boundary of the domain onto the whole domain. The best extension problem is solved here to within an arbitrary $\epsilon > 0$ (see Theorems [5.4] and [8.4] as well as Theorem [12.3]).

In Chapter II we consider spaces of functions having weak derivatives that are summable in some power with a weight; we prove *theorems on the imbedding of these spaces into ordinary function spaces* and investigate completeness, compactness, etc.

Chapter III is devoted to the presentation of general *variational principles concerning the first boundary problem* for selfadjoint elliptic equations of second order, and to the proof of existence and uniqueness theorems for the boundary problems of elliptic differential equations that are degenerate on the boundary of the domain.

A large part of the results of the present monograph has been published in notes of the author [13, 14]; moreover, we make use of the results of the author in [15, 16].

We now consider in some detail the contents of the individual sections.

In §1 we introduce an averaging operator K with a variable averaging radius relative to the hyperplane $x_n = 0$ and prove its boundedness in $L_p^{(n)}$.

In §2 we prove in a sense the invariance under K of the H-classes of Nikol′skiĭ [3].

In §3 we establish the existence of and give formulas for the derivatives of the functions Kf, $f \in L_p^{(n)}$.

In §4 we investigate the behavior of a function Kf and its derivatives on the hyperplane $x_n = 0$. In particular, we prove one of the basic properties of the operator K and similar operators, viz. that *the boundary values of an averagable function f of a certain class coincide with the boundary values of the averaged function.* We also study in this section the connection between the boundary values of the derivatives of functions and their averagings. Here also we present Nikol′skiĭ's results on the rate of convergence in the mean of the functions of an H-class to their boundary values.

In §5 we give an estimate of the growth of the derivatives of a function Kf, after which we prove the main theorem of this part of the book, which asserts that every function given on the hyperplane $x_n = 0$ and belonging to a certain H-class can be extended onto the whole space as an infinitely continuously differentiable function whose normal derivatives on the hyperplane $x_n = 0$ take preassigned values and for which it is possible to indicate a power order of growth for all of its derivatives under an approach to the hyperplane $x_n = 0$. In this connection the *minimal possible growth exponent* is determined *to within an arbitrary* $\epsilon > 0$, i.e. the subsequent results of §12 imply that the order of growth obtained for the derivatives is best possible to within an arbitrary $\epsilon > 0$ (see Theorem [12.3]). Moreover, the constructed extension preserves to within constants all of the properties of the extensions obtained by Nikol′skiĭ [4] for the same class of boundary functions.

The method of proof of this theorem consists in extending a selected auxiliary system of functions by Nikol′skiĭ's method in a certain way and then applying the averaging operator K, the above mentioned properties of which ensure the required properties of the resulting extension of a boundary function.

§6 contains (i) a generalization of a theorem of Nikol′skiĭ on function extensions from a subdomain of a finite domain onto the whole space to the case of an infinite domain, and (ii) a study of function extensions by the method of "symmetry," the idea for which goes back to Whitney and Hestenes.

At the beginning of §7 we give a general method for averaging functions with a variable averaging radius relative to a given m-dimensional hyperplane in n-dimensional space, $1 \leqslant m < n$.

§8 contains a brief sketch of the part of the theory of manifolds that is applicable in the sequel. Also in this section we study the question of the connection between function extensions from flat manifolds and from arbitrary manifolds; in particular, we generalize the results of §5 to the case of arbitrary (not necessarily closed or finite) manifolds. At the end of the section we make some remarks on function extensions from domains bounded by sufficiently smooth manifolds onto the whole space with preservation of the H-class to within constants.

In §9 we consider a domain G in the halfspace $x_n > 0$ that is bounded by a finite number of pieces of $(n-1)$-dimensional manifolds and the hyperplane $x_n = 0$. We give a method of extending functions given on the boundary of G onto the whole domain under which the derivatives of the extension are square summable over G with weight x_n^α, $\alpha > 0$.

§10 contains a direct application of the results. This application is based on the fact that the constructed function extensions substantially improve the properties of the functions being extended. Here we indicate the possibility of applying the above function extensions to the solution of boundary problems for partial differential equations by the method of variation of boundaries; in particular, when in conjunction with a variation of the appropriate integral we consider a variation of the boundary of the domain. In addition to one of the results of Keldyš and Lavrent′ev [1] we indicate here the "rate" of convergence to a solution. This method is illustrated by the well-studied Dirichlet problem for the Laplace equation. It is shown in particular that such a dual variational method can be used to solve the Dirichlet problem for the unit disk in the case of the classical example of Hadamard, when the ordinary variational method does not apply.

In §11 we prove a number of integral inequalities with weight for use in the sequel; in particular, we use them to obtain a theorem on the imbedding into H-classes of spaces of functions whose derivatives of a certain order are summable in some power with weight. Here also we continue the study of the method of "symmetric" extensions (see §6).

§12 is devoted to a generalization of the results of the preceding section to the case of arbitrary manifolds. In addition, we study the connection between the values of a function in a domain and its boundary values for the case when the derivatives of the function are summable in some power with a power weight relative to the distance from a point to the boundary of the domain; we give a sufficient criterion for extending a function from a domain bounded by a smooth manifold onto the whole space with preservation of its H-class to within constants; and finally, we prove an inequality of Poincare type with weight that connects a norm of a function and its weak derivatives in a domain with a norm of the function on the boundary of the domain.

We note that at the present time necessary and sufficient conditions for a function defined on the boundary of a finite domain to be equal to the boundary values of a function whose derivatives are summable with a degenerate weight have been obtained with the use of other methods and in other terms by P. I. Lizorkin [50] (and somewhat earlier for a special case of A. A. Vašarin [49]).

In §13 we prove the completeness of the spaces with weight considered by us.

In §14 we investigate by the method of diagonal sums the connection between the convergence of a sequence of functions in a domain and the convergence of their boundary values.

In §15 we give compactness criteria for both ordinary function spaces and spaces with weight. In particular, the sufficient compactness criteria presented here complement the corresponding results of Kondrašov and Sobolev [2] and Nikol'skiĭ [3]. The complementation is due to (i) an assertion on the simultaneous convergence of a selected subsequence on all cross sections of any compact sets by m-dimensional hyperplanes, the convergence being uniform relative to the choice of hyperplane, (ii) an assumption on the summability of the derivatives not simply raised to some power but on the corresponding summability of them with a certain degenerate weight, and (iii) the consideration of normal derivatives.

The results of §§14 and 15 are based on results published in the author's note [15].

In §16 some of the results of §§11–15 are carried over to the domain with piecewise smooth boundary considered in §9.

In §17, for an arbitrary bounded domain of n-dimensional space, we construct the variational method of solving the first boundary problem for a selfadjoint elliptic equation of second order. The absence of any smoothness conditions for the boundary of the domain is achieved by understanding the boundary values *not in the sense of convergence in the mean*, as is usually done in this case, but *in the sense of convergence "almost everywhere."* In this connection we essentially make use of the results of the author's article [16], which are formulated here. In the same section we present investigations of the smoothness of the resulting solutions, in which the results of Kondrašov play a major role.

It should be noted that in this section, despite a maximal in a sense generality of presentation of the variational principles, under the classical statement of the problem, i.e. when the admissible classes are nonempty, we do not here make use of any of the imbedding theorems of Sobolev, Kondrašov, Nikol′skiĭ and others, but give (as in the article [16], to which we refer) completely elementary proofs of the theorems [17.1] and [17.2] cited by us. An understanding of §17 does not require a knowledge of either the methods or the results presented in §§1–16.

Since we were interested in making the presentation of these questions as simple as possible, we did not seek to prove the theorems obtained in this section under minimal requirements on the coefficients. In certain cases, as is well known, under additional restrictions on the class of domains being considered the conditions imposed by us on the coefficients of the equations can be weakened.

The last four sections are devoted to a study of the equation

$$L_{\sigma^\alpha}(u) = \sum_{i=1}^{n} \frac{\partial}{\partial x_i}\left(\sigma^\alpha \frac{\partial u}{\partial x_i}\right) = 0, \quad \alpha > 0,$$

which is the Euler equation for the functional

$$D_{\sigma^\alpha}(u) = \int \cdots \int \sigma^\alpha \sum_{i=1}^{n} \left(\frac{\partial u}{\partial x_i}\right)^2 dx_1, \ \ldots, \ dx_n, \quad \sigma > 0.$$

It is assumed in this connection that σ tends to zero like the distance from a point of the domain to its boundary under an approach to the boundary of the domain.

In the case when $\sigma = x_n$ the equation $L_{\sigma^\alpha}(u) = 0$ takes the form

$$x_n \Delta u + \alpha \frac{\partial u}{\partial x_n} = 0, \ \alpha > 0.$$

As is known, in the case $n = 2$, for example, equations of the form $y^2\partial^2u/\partial x^2 + \partial^2u/\partial y^2 = 0$ reduce to this same equation when $\alpha > 0$ as well as when $\alpha < -2$.

A large number of investigations have been devoted to the study of similar equations both from the point of view of equations of mixed type and from the point of view of equations of elliptic type in the upper halfplane. Of these we mention the works of Tricomi [18], Lavrent′ev and Bicadze [19], Bicadze [20], Keldyš [21], Vekua [22], Višik [23], Mihlin [24], Huber [25], Germain and Bader [26] and Vašarin [49]. Further references can be found in the bibliographies of these papers.

The most complete investigation of boundary problems for general degenerate elliptic equations is given in a paper of Višik [23].

In §18 we give a general plan for solving the equation $L_{\sigma\alpha}(u) = 0$ by the variational method. We show that it leads to a representation of the desired solution in the form $u = \Phi + \Psi$, where Φ is an averaging of u and Ψ is an integral of Newtonian potential type with mass a linear function of u.

The connection between the smoothness of the function σ and the smoothness of the resulting solution is studied on the basis of the results of Sobolev and Kondrašov [2, 10] on integrals of potential type and certain additional investigations.

In §19 we consider the case of a degeneracy of elliptic equations on the whole boundary of a finite domain and prove an existence and uniqueness theorem for the solution of the boundary problem. We also examine the case of a halfspace studied earlier by Vekua [22] from another point of view and under other assumptions.

In §20 we prove an existence and uniqueness theorem for the solution of a boundary problem when the degeneracy of an elliptic equation occurs on part of the boundary of the domain. The cases of a finite domain and a strip are investigated.

It is assumed in both §§19 and 20 that the degeneracy exponent $\alpha \in [0, 1)$; the case $\alpha \geqslant 1$ is studied in §21.

In conclusion we note that for an understanding of the basic contents of the present monograph one need only be familiar with the first sections of the monograph of Sobolev [2], two papers of Nikol′skiĭ [3, 4] and the author's articles [15, 16].

I wish to take this opportunity to express my sincere thanks to V. I. Kondrašov and P. I. Lizorkin for reading the manuscript of the present monograph and making a number of valuable remarks, all of which were taken into account in arriving at the final wording of the text.

First Chapter

FUNCTION EXTENSIONS

§1. The averaging operator K

Let $L_p^{(n)}$ denote the space of functions that are p power summable on all of the n-dimensional space E^n, $n = 1, 2, \cdots$, and let $\|f\|_p^{(n)}$ denote the norm of a function $f \in L_p^{(n)}$ in the sense of $L_p^{(n)}$. We will always assume that $1 \leqslant p \leqslant \infty$, it being understood that by $\|f\|_\infty^{(n)}$ is meant, as usual, ess $\sup_{x \in E^n} |f(x)|$.

Consider the functions $K(t) = e^{-1/(1-t^2)}$ for $|t| \leqslant 1$ and $K(u, x, \xi) = K(2(\xi - x)/u) = \exp(-u^2/(-u^2 - 4(x - \xi)^2))$ for $|\xi - x| \leqslant |u|/2, -\infty < u < \infty$. We note the following obvious formulas:

$$0 \leqslant K(u, x, \xi) \leqslant \frac{1}{e} \quad \text{for} \quad |\xi - x| \leqslant \frac{|u|}{2}; \tag{1.1}$$

$$\left| \frac{\partial^s K(u, x, \xi)}{\partial u^{s_1} \partial x^{s_2} \partial \xi^{s_3}} \right| \leqslant \frac{C_s}{|u|^s} \quad \text{for } s_1 + s_2 + s_3 = s, \quad |\xi - x| \leqslant \frac{|u|}{2}, \tag{1.2}$$

where the constants C_s depend only on $s = 1, 2, \cdots$; and

$$\frac{\partial^s K\left(u, x, x \pm \frac{u}{2}\right)}{\partial u^{s_1} \partial x^{s_2} \partial \xi^{s_3}} = 0 \quad \text{for } s_1 + s_2 + s_3 = s, \; s = 0, 1, 2, \ldots . \tag{1.3}$$

We next put

$$K(x, \xi) = K(x_1, \ldots, x_n, \xi_1, \ldots, \xi_n) = \prod_{i=1}^{n} K(x_n, x_i, \xi_i), \tag{1.4}$$

$$c = \frac{1}{2^n} \left[\int_{-1}^{1} K(t)\, dt \right]^n . \tag{1.5}$$

Then

$$\frac{1}{c x_n^n} \int_{x_1 - \frac{x_n}{2}}^{x_1 + \frac{x_n}{2}} \cdots \int_{x_n - \frac{x_n}{2}}^{x_n + \frac{x_n}{2}} K(x_1, \ldots, x_n, \xi_1, \ldots, \xi_n)\, d\xi_1 \ldots d\xi_n = 1. \tag{1.6}$$

For, making the change of variable

$$\xi_i = x_i + \frac{x_n t_i}{2}, \quad i = 1, 2, \ldots, n; \tag{1.7}$$

we get

$$\frac{1}{cx_n^n} \int_{x_1 - \frac{x_n}{2}}^{x_1 + \frac{x_n}{2}} \cdots \int_{x_n - \frac{x_n}{2}}^{x_n + \frac{x_n}{2}} K(x_1, \ldots, x_n, \xi_1, \ldots, \xi_n)\, d\xi_1 \ldots d\xi_n$$

$$= \frac{1}{2^n c} \int_{-1}^{1} \cdots \int_{-1}^{1} \prod_{i=1}^{n} K(t_i)\, dt_1 \ldots dt_n = \frac{1}{2^n c} \prod_{i=1}^{n} \int_{-1}^{1} K(t_i)\, dt_i = 1.$$

We now define the integral transform K for any function $f \in L_p^{(n)}$ as follows:

$$Kf = Kf(x_1, \ldots, x_n)$$
$$= \frac{1}{cx_n^n} \int_{x_1 - \frac{x_n}{2}}^{x_1 + \frac{x_n}{2}} \cdots \int_{x_n - \frac{x_n}{2}}^{x_n + \frac{x_n}{2}} K(x_1, \ldots, x_n, \xi_1, \ldots, \xi_n) f(\xi_1, \ldots, \xi_n) d\xi_1 \ldots d\xi_n. \text{(1)} \tag{1.8}$$

By means of the change of variable (1.7) we can express this transform in the form

$$Kf = \frac{1}{2^n c} \int_{-1}^{1} \cdots \int_{-1}^{1} \prod_{i=1}^{n} K(t_i) f\left(x_1 + \frac{x_n t_1}{2}, \ldots, x_n + \frac{x_n t_n}{2}\right) dt_1 \ldots dt_n. \tag{1.9}$$

Clearly, K is an averaging operator with a variable averaging "radius", and the value of the function $Kf(x_1, \cdots, x_n)$ is finite at any point $(x_1, \cdots, x_n) \in E^n, x_n \neq 0$, for any $f \in L_p^{(n)}$. For the sake of brevity we will use the following notation:

$$K^{(i)} = K(x_n, x_i, \xi_i), \quad i = 1, 2, \ldots, n,$$
$$Q_{x_n} = \left\{(\xi_1, \ldots, \xi_n) : |\xi_i - x_i| \leqslant \frac{|x_n|}{2}, \ i = 1, 2, \ldots, n\right\}, \tag{1.10}$$
$$x = (x_i) = (x_1, \ldots, x_n) \in E^n;$$

in this notation we have, for example,

$$Kf(x) = \frac{1}{cx_n^n} \int \prod_{i=1}^{n} K^{(i)} f dQ_{x_n}. \tag{1.11}$$

(1) Here and below we consider Kf as a single function symbol.

[1.1]. *For any function* $f \in L_p^{(n)}$

$$\| Kf \|_p^{(n)} \leqslant 2^{\frac{1}{p}} \| f \|^{(n)} \tag{1.12}$$

COROLLARY. *If* $f \in L_p^{(n)}$, *then* $Kf \in L_p^{(n)}$.

PROOF. Suppose $p = \infty$. Then from (1.9) we have

$$| Kf(x) | \leqslant \frac{1}{2^n c} \int_{-1}^{1} \cdots \int_{-1}^{1} \prod_{i=1}^{n} K(t_i) \left| f\left(x_1 + \frac{x_n t_1}{2}, \ldots, x_n + \frac{x_n t_n}{2}\right) \right| dt_1 \ldots dt_n$$
$$\leqslant \frac{\| f \|_\infty^{(n)}}{2^n c} \int_{-1}^{1} \cdots \int_{-1}^{1} \prod K(t_i)\, dt_1 \ldots dt_n = \| f \|_\infty^{(n)},$$

which implies (1.12) for $p = \infty$.

Suppose now $1 \leqslant p < \infty$. In the following proof and repeatedly in the sequel we will apply the so-called generalized Minkowski inequality

$$\left(\int \left| \int F(u, v)\, du \right|^p dv\right)^{\frac{1}{p}} \leqslant \int \left(\int | F(u, v) |^p dv\right)^{\frac{1}{p}} du;$$

here (u, v) is a point of some set E lying in the direct sum $E_{u,v}^{n_1+n_2}$ of Euclidean spaces $E_u^{n_1}$ and $E_v^{n_2}$. We have

$$\| Kf \|_p^{(n)} = \left[\int_{-\infty}^{\infty} \cdots \int_{-\infty}^{\infty}\right] \frac{1}{2^n c} \int_{-1}^{1} \cdots \int_{-1}^{1} \prod_{i=1}^{n} K(t_i) f\left(x_1 + \frac{x_n t_1}{2}, \ldots, x_n + \frac{x_n t_n}{2}\right)$$
$$\times dt_1 \ldots dt_n \Bigg|^p dx_1 \ldots dx_n \Bigg]^{\frac{1}{p}}$$
$$\leqslant \frac{1}{2^n c} \int_{-1}^{1} \cdots \int_{-1}^{1} \prod_{i=1}^{n} K(t_i) \left[\int_{-\infty}^{\infty} \cdots \int_{-\infty}^{\infty} \left| f\left(x_1 + \frac{x_n t_1}{2}, \ldots, x_n + \frac{x_n t_n}{2}\right)\right|^p\right.$$
$$\left. \times dx_1 \ldots dx_n \right]^{\frac{1}{p}} dt_1 \ldots dt_n.$$

In the inner integral we now make the change of variable

$$u_i = x_i + \frac{x_n t_i}{2}, \quad i = 1, 2, \ldots, n. \tag{1.13}$$

This is a nondegenerate linear mapping of E^n onto itself with determinant

$$\frac{\partial (u_1, \dots, u_n)}{\partial (x_1, \dots, x_n)} = 1 + \frac{t_n}{2},$$

so that

$$\left| \frac{\partial (x_1, \dots, x_n)}{\partial (u_1, \dots, u_n)} \right| = \frac{1}{\left| 1 + \frac{t_n}{2} \right|} \leqslant 2 \quad \text{for } |t_n| \leqslant 1, \tag{1.14}$$

and therefore

$$\| Kf \|_p^{(n)} \leqslant \frac{1}{2^n c} \int_{-1}^{1} \dots \int_{-1}^{1} \prod_{i=1}^{n} K(t_i) \left[\int_{-\infty}^{\infty} \dots \int_{-\infty}^{\infty} | f(u_1, \dots, u_n)|^p \right.$$

$$\times \left. \left| \frac{\partial (x_1, \dots, x_n)}{\partial (u_1, \dots, u_n)} \right| du_1 \dots du_n \right]^{\frac{1}{p}} dt_1 \dots dt_n$$

$$\leqslant \frac{1}{2^{n-\frac{1}{p}} c} \| f \|_p^{(n)} \int_{-1}^{1} \dots \int_{-1}^{1} \prod_{i=1}^{n} K(t_i)\, dt_1 \dots dt_n = 2^{\frac{1}{p}} \| f \|_p^{(n)}.$$

We have made use here of equality (1.5).

[1.2]. *K is a continuous linear operator in $L_p^{(n)}$.*

The linearity of K is obvious, while its continuity follows from its boundedness proved above.

§2. The invariance of the H-classes under K

Let $r > 0$. In the sequel $\bar{r}$ and α will always denote the two numbers such that $r = \bar{r} + \alpha$, $\bar{r}$ is a nonnegative integer and $0 < \alpha \leqslant 1$.

$\Delta_{x_i}^{(k)}(f, h)$, as usual, is the difference of order k with step h of a function f with respect to a variable x_i.

By $H_{p(n)}^{(r_1, \dots, r_n)}(M_1, \cdots, M_n)$ we denote the set of all functions $f \in L_p^{(n)}$ having weak derivatives $\partial^{\bar{r}_i} f / \partial x_i^{\bar{r}_i} \in L_p^{(n)}$ (see [2, 3]) such that

$$\left\| \Delta_{x_i}^{(2)} \left(\frac{\partial^{\bar{r}_i} f}{\partial x_i^{\bar{r}_i}}, h \right) \right\|_p^{(n)} \leqslant M_i |h|^{\alpha_i}, \tag{2.1}$$

the M_i being positive constants, $i = 1, \cdots, n$.

Here, if in the case $\alpha_i < 1$ the second difference is replaced by the first, i.e. (2.1) is replaced by the requirement

$$\left\| \Delta_{x_i}^{(1)} \left(\frac{\partial^{\overline{r_i}} f}{\partial x_i^{\overline{r_i}}}, h \right) \right\|_p^{(n)} \leqslant M_i |h|^{\alpha_i}, \tag{2.2}$$

we obtain a definition equivalent to the one given above. In place of $H_{p(n)}^{(r,\cdots,r)}(M, \cdots, M)$ we will simply write $H_{p(n)}^{(r)}(M)$. All of these definitions are due to Nikol'skiĭ [3].

In the present section we show that the H-classes are in a sense left invariant under the integral transform K. To this end we need some properties of entire functions. As is well known, for every entire function of exponential type $g_{\nu_1 \ldots \nu_n}(z_1, \cdots, z_n)$ of degrees ν_i respectively in the variables z_i $(i = 1, \cdots, n)$ and for any $\epsilon > 0$ there exists a constant $A > 0$ such that for all complex $z_1, \cdots, z_n$

$$g_{\nu_1 \ldots \nu_n}(z_1, \ldots, z_n) | < Ae^{\sum_{i=1}^{n} (\nu_i + \epsilon) |z|} \tag{2.3}$$

Let us prove the following lemma.

[2.1.1]. *In order for the entire function*

$$g(z_1, \ldots, z_n) = \sum_{k_1=0}^{\infty} \cdots \sum_{k_n=0}^{\infty} a_{k_1,\ldots,k_n} z_1^{k_1} \cdots z_n^{k_n} \tag{2.4}$$

to be an entire function of exponential type of degrees ν_i *respectively in the variables* z_i $(i = 1, \cdots, n)$ *it is necessary and sufficient that for any* $\epsilon > 0$ *there exist a constant* $B > 0$ *such that for all complex* $z_1, \cdots, z_n$

$$\sum_{k_1=0}^{\infty} \cdots \sum_{k_n=0}^{\infty} |a_{k_1 \ldots k_n}| |z_1|^{k_1} \cdots |z_n|^{k_n} < Be^{\sum_{i=1}^{n} (\nu_i + \epsilon) |z_i|} \tag{2.5}$$

We note that when $n = 1$ this readily follows from the well-known formulas connecting the degree of a function with the coefficients of its power series. The sufficiency of condition (2.5) is obvious; let us prove its necessity for any n.

According to the Cauchy inequality for the coefficients of the power series (2.4) for analytic functions of several variables [27], we have in a polycylinder $|z_i| \leqslant R_i$, $i = 1, \cdots, n$,

$$|a_{k_1 \ldots k_n}| \leqslant \frac{M(R_1, \ldots, R_n)}{R_1^{k_1} \cdots R_n^{k_n}},$$

where $M(R_1, \cdots, R_n)$ is the greatest value of $|g(z_1, \cdots, z_n)|$ in the indicated polycylinder. Hence from (2.3), replacing ϵ by $\epsilon/2$, we get

$$|a_{k_1\dots k_n}| \leqslant \frac{Ae^{\sum_{i=1}^{n}\left(\nu_i+\frac{\varepsilon}{2}\right)R_i}}{R_1^{k_1}\dots R_n^{k_n}}.$$

Put $R_i = k_i/(\nu_i + \epsilon/2)$. Then

$$|a_{k_1\dots k_n}| \leqslant A\prod_{i=1}^{n}\frac{e^{k_i}\left(\nu_i+\frac{\varepsilon}{2}\right)^{k_i}}{k_i^{k_i}}.$$

Noting that $e^m/m^m < (2m+1)/m!$ for any natural m, we have

$$|a_{k_1\dots k_n}| \leqslant A\prod_{i=1}^{n}\frac{(2k_i+1)\left(\nu_i+\frac{\varepsilon}{2}\right)^{k_i}}{k_i!},$$

which implies

$$|a_{k_1\dots k_n}| \leqslant A\prod_{i=1}^{n}\frac{\left(\nu_i+\frac{\varepsilon}{2}\right)^{k_i}}{(k_i-2)!}, \quad k_i = 4, 5, \dots, \; i = 1, 2, \dots, n.$$

Now

$$\sum_{k_1=4}^{\infty}\dots\sum_{k_n=4}^{\infty}|a_{k_1\dots k_n}||z_1|^{k_1}\dots|z_n|^{k_n} \leqslant \sum_{k_1=4}^{\infty}\dots\sum_{k_n=4}^{\infty}\prod_{i=1}^{n}\frac{\left(\nu_i+\frac{\varepsilon}{2}\right)^{k_i}}{(k_i-2)!}|z_i|^{k_i}$$

$$= A\left[\prod_{i=1}^{n}\left(\nu_i+\frac{\varepsilon}{2}\right)^2|z_i|^2\right]\prod_{i=1}^{n}\sum_{k_i=0}^{\infty}\frac{\left(\nu_i+\frac{\varepsilon}{2}\right)^{k_i}}{k_i!}|z_i|^{k_i}$$

$$= A\prod_{i=1}^{n}\left(\nu_i+\frac{\varepsilon}{2}\right)^2|z_i|^2 e^{\sum_{j=1}^{n}\left(\nu_j+\frac{\varepsilon}{2}\right)|z_j|}$$

Further, there exists a constant $B_1 > 0$ such that

$$A\prod_{i=1}^{n}\left(\nu_i+\frac{\varepsilon}{2}\right)^2|z_i|^2 \leqslant B_1 e^{\sum_{j=1}^{n}\frac{\varepsilon}{2}|z_j|}$$

Therefore

$$\sum_{k_1=4}^{\infty}\dots\sum_{k_n=4}^{\infty}|a_{k_1\dots k_n}||z_1|^{k_1}\dots|z_n|^{k_n} \leqslant B_1 e^{\sum_{j=1}^{n}(\nu_j+\varepsilon)|z_j|}$$

On the other hand, there obviously exists a constant $B_2 > 0$ for which

$$\sum_{k_1=0}^{3} \cdots \sum_{k_n=0}^{3} |a_{k_1 \ldots k_n}| |z_1|^{k_1} \cdots |z_n|^{k_n} \leqslant B_2 e^{\sum_{j=1}^{n} (\nu_j + \varepsilon)|z_j|}.$$

Adding the last two inequalities, we obtain the desired inequality (2.5). The lemma is proved.

We note that formula (1.9) permits one to apply the operator K to functions defined for all complex values of the arguments.

[2.1]. *Suppose*

$$f(z_1, \ldots, z_n) = \sum_{k_1=0}^{\infty} \cdots \sum_{k_n=0}^{\infty} a_{k_1 \ldots k_n} z_1^{k_1} \cdots z_n^{k_n} \tag{2.6}$$

is an entire function. Then Kf is also an entire function. If f is of exponential type of degree $\nu_1, \cdots, \nu_n$ respectively in the variables $z_1, \cdots, z_n$, then Kf is also of exponential type of degrees $\nu_1, \cdots, \nu_{n-1}, \nu_n + (\nu_1 + \cdots + \nu_n)/2$ respectively in the variables $z_1, \cdots, z_n$.

PROOF. According to (1.9) and (2.6) we have

$$Kf(z_1, \ldots, z_n) = \frac{1}{2^n c} \int_{-1}^{1} \cdots \int_{-1}^{1} \prod_{i=1}^{n} K(t_i) \sum_{k_1=0}^{\infty} \cdots \sum_{k_n=0}^{\infty} a_{k_1 \ldots k_n} \left(z_1 + \frac{z_n t_1}{2}\right)^{k_1} \cdots$$

$$\cdots \left(z_n + \frac{z_n t_n}{2}\right)^{k_n} dt_1 \ldots dt_n$$

$$= \frac{1}{2^n c} \sum_{k_1=0}^{\infty} \cdots \sum_{k_n=0}^{\infty} a_{k_1 \ldots k_n} \int_{-1}^{1} \cdots \int_{-1}^{1} \prod_{i=1}^{n} K(t_i) \prod_{j=1}^{n} \sum_{s_j=0}^{k_j} C_{k_j}^{s_j} z_j^{s_j} \left(\frac{z_n}{2}\right)^{k_j - s_j} t_j^{k_j - s_j} dt_1 \ldots dt_n$$

$$= \frac{1}{2^n c} \sum_{k_1=0}^{\infty} \cdots \sum_{k_n=0}^{\infty} a_{k_1 \ldots k_n} \sum_{s_1=0}^{k_1} \cdots \sum_{s_n=0}^{k_n} \prod_{j=1}^{n} C_{k_j}^{s_j} z_j^{s_j} \left(\frac{z_n}{2}\right)^{k_j - s_j} \int_{-1}^{1} \cdots \int_{-1}^{1} \prod_{i=1}^{n} K(t_i)\, t_j^{k_j - s_j} dt_1 \ldots dt_n$$

$$= \sum_{k_1=0}^{\infty} \cdots \sum_{k_n=0}^{\infty} \sum_{s_1=0}^{k_1} \cdots \sum_{s_n=0}^{k_n} a_{k_1 \ldots k_n} b_{k_1 \ldots k_n s_1 \ldots s_n} \prod_{j=1}^{n} C_{k_j}^{s_j} z_j^{s_j} \left(\frac{z_n}{2}\right)^{k_j - s_j}, \tag{2.7}$$

where

$$b_{k_1 \ldots k_n s_1 \ldots s_n} = \frac{1}{2^n c} \int_{-1}^{1} \cdots \int_{-1}^{1} \prod_{i=1}^{n} K(t_i) \prod_{j=1}^{n} t_j^{k_j - s_j} dt_1 \ldots dt_n,$$

it following from (1.5) that

$$|b_{k_1\ldots k_n s_1\ldots s_n}| \leqslant 1. \tag{2.8}$$

Let us show that the series (2.7) absolutely converges for all complex values of the arguments $z_1, \cdots, z_n$, i.e. that Kf is an entire function. To this end we consider the series

$$F(z_1, \ldots, z_n) = \sum_{k_1=0}^{\infty} \cdots \sum_{k_n=0}^{\infty} a_{k_1\ldots k_n}\left(z_1 + \frac{z_n}{2}\right)^{k_1} \cdots \left(z_n + \frac{z_n}{2}\right)^{k_n}. \tag{2.9}$$

It is obvious that this series converges for any $z_1, \cdots, z_n$ to an entire function $F(z_1, \cdots, z_n)$. Removing the parentheses in (2.9), we again obtain a series that converges absolutely for all complex $z_1, \cdots, z_n$:

$$\begin{aligned} F(z_1, \ldots, z_n) &= \sum_{k_1=0}^{\infty} \cdots \sum_{k_n=0}^{\infty} a_{k_1\ldots k_n} \prod_{j=1}^{n} \sum_{s_j=0}^{k_j} C_{k_j}^{s_j} z_j^{s_j} \left(\frac{z_n}{2}\right)^{k_j - s_j} \\ &= \sum_{k_1=0}^{\infty} \cdots \sum_{k_n=0}^{\infty} \sum_{s_1=0}^{\infty} \cdots \sum_{s_n=0}^{\infty} a_{k_1\ldots k_n} \prod_{j=1}^{n} C_{k_j}^{s_j} z_j^{s_j} \left(\frac{z_n}{2}\right)^{k_j - s_j} \end{aligned} \tag{2.10}$$

But by virtue of (2.8) the absolute values of the terms of this series majorize the absolute values of the terms of (2.7). Hence the series (2.7) converges absolutely for all complex $z_1, \cdots, z_n$ if the series (2.6) converges for them.

Suppose, finally, (2.6) is of exponential type of degrees $\nu_1, \cdots, \nu_n$ respectively in the variables $z_1, \cdots, z_n$. Then, according to [2.1.1], for any $\epsilon > 0$ there exists a constant $B > 0$ such that for any complex $\zeta_1, \cdots, \zeta_n$

$$\sum_{k_1=0}^{\infty} \cdots \sum_{k_n=0}^{\infty} |a_{k_1\ldots k_n}| |\zeta_1|^{k_1} \cdots |\zeta_n|^{k_n} \leqslant B e^{\sum_{i=1}^{n}\left(\nu_i + \frac{\varepsilon}{n+1}\right)|\zeta_i|}$$

We next choose arbitrary complex $z_1, \cdots, z_n$ and put

$$\zeta_i = |z_i| + \frac{|z_n|}{2}, \quad i = 1, 2, \ldots, n.$$

Then

$$\sum_{k_1=0}^{\infty}\cdots\sum_{k_n=0}^{\infty}|a_{k_1\ldots k_n}|\prod_{i=1}^{n}\left(|z_i|+\frac{|z_n|}{2}\right)^{k_i}\leqslant Be^{\sum_{i=1}^{n}\left(\nu_i+\frac{\varepsilon}{n+1}\right)\left(|z_i|+\frac{|z_n|}{2}\right)}$$

$$\leqslant Be^{\sum_{i=1}^{n-1}(\nu_i+\varepsilon)|z_i|+\left(\frac{\nu_1+\ldots+\nu_n}{2}+\nu_n+\varepsilon\right)|z_n|} \tag{2.11}$$

Now, using (2.7), (2.8) and (2.11), we have

$$|Kf(z_1,\ldots,z_n)|\leqslant\sum_{k_1=0}^{\infty}\cdots\sum_{k_n=0}^{\infty}\sum_{s_1=0}^{k_1}\cdots\sum_{s_n=0}^{k_n}|a_{k_1\ldots k_n}||b_{k_1\ldots k_n s_1\ldots s_n}|\prod_{j=1}^{n}C_{k_j}^{s_j}|z_j|^{s_j}\left|\frac{z_n}{2}\right|^{k_j-s_j}$$

$$\leqslant\sum_{k_1=0}^{\infty}\cdots\sum_{k_n=0}^{\infty}|a_{k_1\ldots k_n}|\prod_{j=1}^{n}\sum_{s_j=0}^{k_j}C_{k_j}^{s_j}|z_j|^{s_j}\left|\frac{z_n}{2}\right|^{k_j-s_j}$$

$$=\sum_{k_1=0}^{\infty}\cdots\sum_{k_n=0}^{\infty}|a_{k_1\ldots k_n}|\prod_{j=1}^{n}\left(|z_j|+\frac{|z_n|}{2}\right)^{k_j}$$

$$\leqslant Be^{\sum_{i=1}^{n-1}(\nu_i+\varepsilon)|z_i|+\left(\frac{\nu_1+\ldots+\nu_n}{2}+\nu_n+\varepsilon\right)|z_n|}$$

for the arbitrarily taken $z_1,\cdots,z_n$, and this means that $Kf(z_1,\cdots,z_n)$ is a function of exponential type of corresponding degrees in its arguments.

REMARK. In the case $n=1$ the first part of Theorem [2.1] can be sharpened: if

$$f(z)=\sum_{0}^{\infty}a_nz^n \tag{2.12}$$

is an analytic function in the disk $|z|<R$ then Kf is an analytic function in the disk $|z|<R/2$, it being true that if $Kf(z)=\Sigma_0^\infty b_nz^n$, then

$$|b_n|\leqslant 2^n|a_n|,\quad n=0,\ 1,\ \ldots. \tag{2.13}$$

For, when $n=1$,

$$Kf(z)=\int_{-1}^{1}K(t)f\left[z\left(1+\frac{t}{2}\right)\right]dt.$$

Substituting (2.12), we have

$$Kf(z) = \frac{1}{2c}\sum_0^\infty a_n z^n \int_{-1}^{1} K(t)\left(1+\frac{t}{2}\right)^n dt = \sum_0^\infty b_n z^n,$$

where

$$b_n = \frac{a_n}{2c}\int_{-1}^{1} K(t)\left(1+\frac{t}{2}\right)^n dt.$$

But by virtue of (1.5) this implies (2.13) and hence the complete assertion taken as a whole.

[2.2]. *Suppose* $f \in H_{p(n)}^{(r_1,\cdots,r_n)}(M_1,\cdots,M_n)$. *Then*

$$Kf \in H_{p(n)}^{(r_1,\ldots,r_{n-1},r)}(M,\ \ldots,\ M),\ \textit{where}\ r = \min_{i=1,2\ldots n} r_i, \tag{2.14}$$

$$M \leqslant c' \sum_{k=1}^{n} M_k + c'' \| f \|_p^{(n)},$$

the constants c' *and* c'' *not depending on the* M_i *and* f, $i = 1, \cdots, n$.

Corollary. *Suppose* $f \in H_{p(n)}^{(r)}(M)$. *Then* $Kf \in H_{p(n)}^{(r)}(M')$, *where* $M' \leqslant c_1 M + c_2 \| f \|_p^{(n)}$

Proof. Following Nikol'skiĭ [3], we put for a given function f

$$A_{\nu_1\ldots\nu_n}(f)_p = \inf_{g_{\nu_1\ldots\nu_n}} \| f - g_{\nu_1\ldots\nu_n} \|_p^{(n)},$$

$$A_{\nu_j}(f)_p = \inf_{g_{\nu_j}} \| f - g_{\nu_j} \|_p^{(n)}, \quad j = 1,\ 2,\ \ldots,\ n;$$

i.e. $A_{\nu_1\cdots\nu_n}(f)_p$ and $A_{\nu_j}(f)_p$ are the measures of best approximation of f by entire functions of exponential type of degrees $\nu_1, \cdots, \nu_n$ in the variables $x_1, \ldots, x_n$ and of degree ν_j respectively.

Suppose now $f \in H_{p(n)}^{(r_1,\cdots,r_n)}(M_1,\cdots,M_n)$ and g_s is an entire function of degrees $\nu_i = 2^{s/r_i}$ in the variables x_i, $i = 1, \cdots, n$, which effects a best approximation of f in this class, i.e.

$$A_{2^{\frac{s}{r_1}}\ldots 2^{\frac{s}{r_n}}}(f)_p = \| f - g_s \|_p^{(n)}, \quad s = 0,\ 1,\ \ldots.$$

According to a result of Nikol'skiĭ ([3], Theorem 7, page 18),

$$A_{2^{\frac{s}{r_1}}\ldots 2^{\frac{s}{r_n}}}(f)_p \leqslant \frac{d\sum_{i=1}^{n} M_i}{2^s},$$

where d does not depend on f or the M_i. We can therefore expand f in a series that converges in $L_p^{(n)}$:

$$f = \sum_{s=0}^{\infty} q_s, \tag{2.15}$$

where $q_0 = g_0$ and $q_s - g_{s-1}$, $s = 1, 2, \cdots$. Clearly, q_s is an entire function of exponential type of degree $2^{s/r_i}$ in x_i, $i = 1, \cdots, n$, and

$$\| q_0 \|_p^{(n)} = \| g_0 \|_p^{(n)} \leqslant \| f - g_0 \|_p^{(n)} + \| f \|_p^{(n)} \leqslant d \sum_1^n M_i + \| f \|_p^{(n)}, \tag{2.16}$$

$$\| q_s \|_p^{(n)} \leqslant \| g_s - f \|_p^{(n)} + \| f - g_{s-1} \|_p^{(n)} \leqslant \frac{4d\sum_1^n M_i}{2^s}, \quad s = 1, 2, \ldots. \tag{2.17}$$

The constant d does not depend on f or the M_i. We note that the expansion (2.15) with the corresponding estimates was also obtained by Nikol'skiĭ ([3], page 28).

By virtue of the linearity and continuity in $L_p^{(n)}$ of the averaging operator K we obtain from (2.13) the convergent in $L_p^{(n)}$ series

$$Kf = \sum_{s=0}^{\infty} Kq_s. \tag{2.18}$$

According to Theorem [2.1] the function Kq_s is an entire function of exponential type of degree $\nu_i^{(s)} = 2^{s/r_i}$ in x_i, $i = 1, \cdots, n-1$, and of degree $\nu_n^{(s)} = 2^{n+s/r}$ in x_n, where $r = \min_{i=1,\ldots,n} r_i$, since

$$\frac{1}{2}\sum_{i=1}^{n} 2^{\frac{s}{r_i}} + 2^{\frac{s}{r_n}} \leqslant \left(\frac{n}{2}+1\right) 2^{\frac{s}{r}} \leqslant 2^{n+\frac{s}{r}}$$

We now put $Q_\mu = \sum_{s=0}^{\mu} Kq_s$, $\mu = 0, 1, \ldots$. Clearly, Q_μ is also an entire function of degrees $\nu_i^{(\mu)}$ in the variables x_i, $i = 1, \cdots, n$. Applying successively (2.18), [1.1], (2.16) and (2.17), we get

$$\| Kf - Q_\mu \|_p^{(n)} \leqslant \sum_{s=\mu+1}^{\infty} \| Kq_s \|_p^{(n)} \leqslant 2^{\frac{1}{p}} \sum_{s=\mu+1}^{\infty} \| q_s \|_p^{(n)}$$

$$\leqslant 2^{\frac{1}{p}} \sum_{s=\mu+1}^{\infty} \frac{4d \sum_{i=1}^{n} M_i + \| f \|_p^{(n)}}{2^s} = \frac{B}{2^\mu}, \quad \mu = 0, 1, \ldots,$$

where

$$B = 2^{\frac{1}{p}} \left(4d \sum_{i=1}^{n} M_i + \| f \|_p^{(n)} \right), \tag{2.19}$$

which implies

$$\| Kf - Q_\mu \|_p^{(n)} \leqslant \frac{B}{(\nu_i^{(\mu)})^{r_i}}, \quad i = 1, 2, \ldots, n-1, \tag{2.20}$$

$$\| Kf - Q_\mu \|_p^{(n)} \leqslant \frac{2^{r_n} B}{(\nu_n^{(\mu)})^{r}}. \tag{2.21}$$

By a theorem of Nikol′skiĭ ([3], Theorem 8, page 21), if for a function $F \in L_p^{(n)}$ the estimate $A_{\nu_j}(F)_p \leqslant K/\nu_j^\rho$ holds for all ν_j ranging over a geometric progression with common ratio greater than one,(2) then

$$F \in H_{px_j}^{(\rho)}(M), \quad M \leqslant c'K + c'' \| F \|_p^{(n)},$$

where c' and c'' do not depend on F or K.

Applying this theorem to the function Kf we conclude from (2.20) and (2.21) that

$$Kf \in H_{px_i}^{(r_i)}(M_i'), \quad M_i' \leqslant c'B + c'' \| f \|_p^{(n)}, \quad i = 1, 2, \ldots, n-1,$$

$$Kf \in H_{px_n}^{(r)}(M_n'), \quad M_n' \leqslant 2^{r_n} c'B + c'' \| f \|_p^{(n)}.$$

Replacing the constant B here by its expression (2.19), we obtain the desired inequality (2.14). The theorem is proved.

In conclusion we note that Theorem [2.2], which has been obtained here with the use of the theory of best approximations, can also be obtained from the results of §3, which do not depend on the contents of the present section. But this requires more preliminary considerations; besides Theorem [2.1] proved above is of independent interest.

(2) The indicated theorem is formulated in the cited paper only for $\nu_k = a^k$, $a > 1$; but it remains true for the case of progressions of the form $\nu_k = ba^k$, $a > 1$, which we also encounter.

§3. The derivatives of Kf

[3.1]. *Suppose* $f \in L_p^{(n)}$. *Then the function* Kf *has continuous derivatives of all orders throughout the space* E^n *except, perhaps, on the hyperplane* $x_n = 0$.

The derivatives are calculated according to the usual rules of differention, taking into account the boundary values of the averaging kernel (*see* (1.3)); *for example,*

$$\frac{\partial Kf(x)}{\partial x_i} = \frac{1}{cx_n^n}\int \frac{\partial K(x,\xi)}{\partial x_i} f(\xi)\,dQ_{x_n}, \quad i = 1, 2, \ldots, n-1, \tag{3.1}$$

$$\frac{\partial Kf(x)}{\partial x_n} = \frac{1}{cx_n^n}\int \frac{\partial K(x,\xi)}{\partial x_n} f(\xi)\,dQ_{x_n} - \frac{n}{cx_n^{n+1}}\int K(x,\xi) f(\xi)\,dQ_{x_n}. \tag{3.2}$$

We prove, as an example, the existence of $\partial Kf/\partial x_n$. For fixed $x_1, \cdots, x_n, h$ and $0 < |h| < |x_n|$ we have

$$\frac{Kf(x_1,\ldots x_{n-1}, x_n+h) - Kf(x_1,\ldots x_{n-1}, x_n)}{h} = \frac{1}{ch}\left[\frac{1}{(x_n+h)^n} - \frac{1}{x_n^n}\right]$$

$$\times \int\limits_{x_1-\frac{x_n+h}{2}}^{x_1+\frac{x_n+h}{2}} \cdots \int\limits_{x_n+h-\frac{x_n+h}{2}}^{x_n+h+\frac{x_n+h}{2}} \prod_{i=1}^{n} K(x_n+h, x_i+\delta_i^n h, \xi_i) f(\xi_1,\ldots,\xi_n)\,d\xi_1\ldots d\xi_n$$

$$+\frac{1}{chx_n^n}\left[\int\limits_{x_1-\frac{x_n+h}{2}}^{x_1+\frac{x_n+h}{2}} \cdots \int\limits_{x_n+h-\frac{x_n+h}{2}}^{x_n+h+\frac{x_n+h}{2}} \prod_{i=1}^{n} K(x_n+h, x_i+\delta_i^n h, \xi_i) f(\xi_1,\ldots,\xi_n)\,d\xi_1\ldots d\xi_n\right.$$

$$\left. - \int\limits_{x_1-\frac{x_n}{2}}^{x_1+\frac{x_n}{2}} \cdots \int\limits_{x_n-\frac{x_n}{2}}^{x_n+\frac{x_n}{2}} \prod_{i=1}^{n} K(x_n+h, x_i+\delta_i^n h, \xi_i) f(\xi_1,\ldots,\xi_n)\,d\xi_1\ldots d\xi_n\right]$$

$$+\sum_{i=1}^{n}\frac{1}{cx_n^n}\int\limits_{x_1-\frac{x_n}{2}}^{x_1+\frac{x_n}{2}} \cdots \int\limits_{x_n-\frac{x_n}{2}}^{x_n+\frac{x_n}{2}} \frac{K(x_n+h, x_i+\delta_i^n h, \xi_i) - K(x_n, x_i, \xi_i)}{h}$$

$$\times \prod_{j=1}^{i-1} K(x_n+h, x_j+\delta_j^n h, \xi_j) \prod_{k=i+1}^{n} K(x_n, x_k, \xi_k) f(\xi_1,\ldots,\xi_n)\,d\xi_1,\ldots,d\xi_n. \tag{3.3}$$

Here δ_i^n is the Kronecker delta: $\delta_n^n = 1$ and $\delta_i^n = 0$ for $i \neq n$. The limit of the first summand for $h \to 0$ is equal to

$$-\frac{n}{cx_n^{n+1}}\int K(x,\xi)f(\xi)\,dQ_{x_n}.$$

This follows from the membership of f in $L_p^{(n)}$, the uniform continuity of the kernels $K(x_n, x_i, \xi_i)$ in the cube Q_{x_n} and the fact that when

$$\left|\xi_i-\left(x_i\pm\frac{x_n}{2}\right)\right|\leqslant\frac{|h|}{2}$$

we have the inequality

$$|K(x_n+h,\ x_i+\delta_i^n h,\ \xi_i)|\leqslant\exp\left[-\frac{(|x_n|-|h|)^2}{(|x_n|+|h|)^2-(|x_n|-|h|)^2}\right], \qquad (3.4)$$
$$i=1,\ 2,\ \ldots,\ n,$$

it being obvious that the right side of this inequality tends to zero as $h\to 0$. The second summand of (3.3), by subdividing the domain of integration, can be represented by a sum of integrals in each of which one of the variables of integration ξ_i varies, so that $|\xi_i-(x_i\pm x_n/2)|\leqslant|h|/2$. Therefore estimate (3.4) is also valid here, which implies that the whole second summand tends to zero as $h\to 0$.

Finally, noting that

$$\left|\frac{K(x_n+h,\ x_i+\delta_i^n h,\ \xi_i)-K(x_n,x_i,\xi_i)}{h}-\frac{\partial K(x_n,x_i,\xi_i)}{\partial x_n}\right|$$
$$\leqslant\left|\frac{h}{2}\,\frac{\partial^2 K(x_n+\theta h,\ x_i+\delta_i^n\theta h,\ \xi_i)}{\partial x_n^2}\right|\leqslant K_0 h,$$

where K_0 is a positive constant and $0<\theta<1$, we have

$$\left|\frac{1}{cx_n^n}\int\left[\frac{K(x_n+h,\ x_i+\delta_i^n h,\ \xi_i)-K(x_n,x_i,\xi_i)}{h}-\frac{\partial K(x_n,x_i,\xi_i)}{\partial x_n}\right]\right.$$
$$\times\prod_{j=1}^{i-1}K(x_n+h,\ x_j+\delta_j^n h,\ \xi_j)\prod_{k=i+1}^{n}K(x_n,x_k,\xi_k)$$
$$\left.\times f(\xi_1\ldots,\xi_n)\,d\xi_1\ldots d\xi_n\right|\leqslant$$

$$\leqslant \frac{K_0 |h|}{c |x_n^n|} \left\{ \int \left[\prod_{j=1}^{i-1} K(x_n + h, x_j + \delta_j^n h, \xi_j) \prod_{k=i+1}^{n} K(x_n, x_k, \xi_k) \right]^q dQ_{x_n} \right\}^{\frac{1}{q}}$$

$$\times \left\{ \int |f|^p dQ_{x_n} \right\}^{\frac{1}{p}}, \quad \frac{1}{p} + \frac{1}{q} = 1.$$

Both of the expressions in braces are bounded; the first, as a consequence of the continuity of the kernels $K(x_n, x_i, \xi_i)$, while the second does not exceed $\|f\|_p^{(n)}$. Therefore the entire right side tends to zero as $h \to 0$. Further,

$$\lim_{h \to 0} \frac{1}{cx_n^n} \int \frac{\partial K(x_n, x_i, \xi_i)}{\partial x_n} \prod_{j=1}^{i-1} K(x_n + h, x_j + \delta_j^n h, \xi_j) \prod_{k=i+1}^{n} K(x_n, x_k, \xi_k) f(\xi_1, \ldots, \xi_n) dQ_{x_n}$$

$$= \frac{1}{cx_n^n} \int \frac{\partial K(x_n, x_i, \xi_i)}{\partial x_n} \prod_{j \neq i} K(x_n, x_j, \xi_j) f(\xi_1, \ldots, \xi_n) \, d\xi_1 \ldots d\xi_n.$$

It follows from what has been proved that the limit of the last sum in (3.3) is

$$\sum_{i=1}^{n} \frac{1}{cx_n^n} \int \frac{\partial K(x_n, x_i, \xi_i)}{\partial x_n} \prod_{j \neq i} K(x_n, x_j, \xi_j) f(\xi_1, \ldots, \xi_n) \, dQ_{x_n}$$

$$= \frac{1}{cx_n^n} \int \frac{\partial K(x, \xi)}{\partial x_n} f(\xi) \, dQ_{x_n}.$$

This completes the proof of the existence of $\partial Kf/\partial x_n$ and formula (3.2).

It is easily seen that the existence of the remaining derivatives of first and higher orders can be proved analogously, provided one notes that each already obtained derivative can be expressed in terms of the function f by means of an integral transform whose kernel is bounded and takes zero values of exponential order on the whole boundary of the cube Q_{x_n}; more precisely, the following theorem holds.

[3.2]. *Suppose* $f \in L_p^{(n)}$ *and* $x_n \neq 0$. *Then*

$$\frac{\partial^{s_1 + \ldots + s_n} Kf(x)}{\partial x_1^{s_1} \ldots \partial x_n^{s_n}} = \frac{1}{c} \sum_{k=0}^{s_n} (-1)^k C_{s_n}^k \frac{\prod_{t=0}^{k-1} (n+t)}{x_n^{n+k}} \int \frac{\partial^{s_1 + \ldots + s_n - k} K(x, \xi)}{\partial x_1^{s_1} \ldots \partial x_{n-1}^{s_{n-1}} \partial x_n^{s_n - k}} f(\xi) \, dQ_{x_n}. \tag{3.5}$$

This formula is obviously valid when $s_1 + \cdots + s_n = 0$. Moreover, it has already been proved by us for the case $s_1 + \cdots + s_n = 1$ (see (3.1) and (3.2)). It is obtained in the general case by induction. We find the derivative of (3.5) with respect to x_n:

$$\frac{\partial^{s_1+\ldots+s_n+1} Kf(x)}{\partial x_1^{s_1}\ldots\partial x_{n-1}^{s_{n-1}}\partial x_n^{s_n+1}}$$

$$= \frac{1}{c}\sum_{k=0}^{s_n}(-1)^k C_{s_n}^k \frac{\prod_{t=0}^{k-1}(n+t)}{x_n^{n+k}} \int \frac{\partial^{s_1+\ldots+s_n+1-k}K(x,\xi)}{\partial x_1^{s_1}\ldots\partial x_{n-1}^{s_{n-1}}\partial x_n^{s_n+1-k}} f(\xi)\,dQ_{x_n}$$

$$-\frac{1}{c}\sum_{k=0}^{s_n}(-1)^k C_{s_n}^k \frac{(n+k)\prod_{t=0}^{k-1}(n+t)}{x_n^{n+k+1}} \int \frac{\partial^{s_1+\ldots+s_n-k}K(x,\xi)}{\partial x_1^{s_1}\ldots\partial x_{n-1}^{s_{n-1}}\partial x_n^{s_n-k}} f(\xi)\,dQ_{x_n}$$

$$= \frac{1}{c}\sum_{k=0}^{s_n}(-1)^k C_{s_n}^k \frac{\prod_{t=0}^{k-1}(n+t)}{x_n^{n+k}} \int \frac{\partial^{s_1+\ldots+s_n+1-k}K(x,\xi)}{\partial x_1^{s_1}\ldots\partial x_{n-1}^{s_{n-1}}\partial x_n^{s_n+1-k}} f(\xi)\,dQ_{x_n}$$

$$+\frac{1}{c}\sum_{k=1}^{s_n+1}(-1)^k C_{s_n}^{k-1} \frac{\prod_{t=0}^{k-1}(n+t)}{x_n^{n+k}} \int \frac{\partial^{s_1+\ldots+s_n+1-k}K(x,\xi)}{\partial x_1^{s_1}\ldots\partial x_{n-1}^{s_{n-1}}\partial x_n^{s_n+1-k}} f(\xi)\,dQ_{x_n}$$

$$= \frac{1}{c}\sum_{k=0}^{s_n+1}(-1)^k C_{s_n+1}^k \frac{\prod_{t=0}^{k-1}(n+t)}{x_n^{n+k}} \int \frac{\partial^{s_1+\ldots+s_n+1-k}K(x,\xi)}{\partial x_1^{s_1}\ldots\partial x_{n-1}^{s_{n-1}}\partial x_n^{s_n+1-k}} f(\xi)\,dQ_{x_n}.$$

As to the derivatives with respect to x_i, $i \neq n$, they are found quite simply; for example, when $i = 1$ we have

$$\frac{\partial^{s_1+\ldots+s_n+1} Kf(x)}{\partial x_1^{s_1+1}\partial x_2^{s_2}\ldots\partial x_n^{s_n}} = \frac{1}{c}\sum_{k=0}^{s_n}(-1)^k C_{s_n}^k \frac{\prod_{t=0}^{k-1}(n+t)}{x_n^{n+k}} \int \frac{\partial^{s_1+\ldots+s_n+1-k}K(x,\xi)}{\partial x_1^{s_1+1}\partial x_2^{s_2}\ldots\partial x_n^{s_n-k}} f(\xi)\,dQ_{x_n}.$$

Formula (3.5) is proved.

In the sequel we need formulas for the derivatives of Kf in which derivatives are taken under the integral sign with respect to the variables of integration. To the deduction of these formulas we now turn.

We first note that (see the notation (1.10))

$$K^{(j)}_{x_i} = \frac{8x_n^2(\xi_j - x_j)}{[x_n^2 - 4(\xi_j - x_j)^2]^2}\left[\delta^i_j + \delta^n_i \frac{\xi_j - x_j}{x_n}\right] K^{(j)},$$

$$K^{(j)}_{\xi_j} = -\frac{8x_n^2(\xi_j - x_j)}{[x_n^2 - 4(\xi_j - x_j)^2]^2} K^{(j)}.\text{(3)}$$

These formulas can be verified by a direct calculation. They imply

$$\left(\delta^i_j + \delta^n_i \frac{\xi_j - x_j}{x_n}\right) K^{(j)}_{\xi_j} = -K^{(j)}_{x_i}. \tag{3.6}$$

In particular, when $i = j \neq n$

$$K^{(j)}_{\xi_j} = -K^{(j)}_{x_j}. \tag{3.7}$$

From (3.6) we obtain the following additional formula needed by us:

$$\frac{\delta^n_{i_{s+1}}}{x_n}\prod_{k=1}^{s}\left(\delta^{i_k}_j + \delta^n_{i_k}\frac{\xi_j - x_j}{x_n}\right)K^{(j)} - \frac{\partial}{\partial x_{i_{s+1}}}\left[\prod_{k=1}^{s}\left(\delta^{i_k}_j + \delta^n_{i_k}\frac{\xi_j - x_j}{x_n}\right)K^{(j)}\right]$$
$$= \frac{\partial}{\partial \xi_j}\left[\prod_{k=1}^{s+1}\left(\delta^{i_k}_j + \delta^n_{i_k}\frac{\xi_j - x_j}{x_n}\right)K^{(j)}\right],$$
$$j = 1, 2, \ldots, n;\ i_k = 1, 2, \ldots, n;\ k = 1, 2, \ldots s+1;\ s = 1, 2, \ldots \tag{3.8}$$

We prove it:

$$\frac{\delta^n_{i_{s+1}}}{x_n}\prod_{k=1}^{s}\left(\delta^{i_k}_j + \delta^n_{i_k}\frac{\xi_j - x_j}{x_n}\right)K^{(j)} - \frac{\partial}{\partial x_{i_{s+1}}}\left[\prod_{k=1}^{s}\left(\delta^{i_k}_j + \delta^n_{i_k}\frac{\xi_j - x_j}{x_n}\right)K^{(j)}\right]$$
$$= \frac{\delta^n_{i_{s+1}}}{x_n}\prod_{k=1}^{s}\left(\delta^{i_k}_j + \delta^n_{i_k}\frac{\xi_j - x_j}{x_n}\right)K^{(j)}$$
$$+ \sum_{k_1=1}^{s}\left(\frac{\delta^n_{i_{k_1}}\delta^{i_{s+1}}_j}{x_n} + \delta^n_{i_{k_1}}\delta^n_{i_{s+1}}\frac{\xi_j - x_j}{x_n^2}\right)\prod_{\substack{k=1\\k\neq k_1}}^{n}\left(\delta^{i_k}_j + \delta^n_{i_k}\frac{\xi_j - x_j}{x_n}\right)K^{(j)} -$$

---

(3) The subscripts x_i and ξ_j denote, as usual, differentiation with respect to the variables x_i and ξ_j respectively.

$$-\prod_{k=1}^{s}\left(\delta_j^{i_k}+\delta_{i_k}^n\frac{\xi_j-x_j}{x_n}\right)K_{x_{i_{s+1}}}^{(j)}=\sum_{k_1=1}^{s+1}\frac{\delta_{i_{k_1}}^n}{x_n}\prod_{\substack{k=1\\k\neq k_1}}^{s+1}\left(\delta_j^{i_k}+\delta_{i_k}^n\frac{\xi_j-x_j}{x_n}\right)K^{(j)}$$

$$+\prod_{k=1}^{s+1}\left(\delta_j^{i_k}+\delta_{i_k}^n\frac{\xi_j-x_j}{x_n}\right)K_{\xi_j}^{(j)}=\frac{\partial}{\partial\xi_j}\left[\prod_{k=1}^{s+1}\left(\delta_j^{i_k}+\delta_{ik}^n\frac{\xi_j-x_j}{x_n}\right)K^{(j)}\right].\text{[4]}$$

[3.3]. *Suppose* $f\in L_p^{(n)}$ *and* $x_n\neq 0$. *Then*

$$\frac{\partial^\lambda Kf(x)}{\partial x_{i_1}\dots\partial x_{i_\lambda}}=\frac{(-1)^\lambda}{cx_n^n}\sum_{\lambda_1+\dots+\lambda_n=\lambda}\int\prod_{j=1}^n\frac{\partial^{\lambda_j}}{\partial\xi_j^{\lambda_j}}\left[\prod_{\nu_j}^{\lambda_j}\left(\delta_j^{i_{\nu_j}}+\delta_{i_{\nu_j}}^n\frac{\xi_j-x_j}{x_n}\right)K^{(j)}\right]fdQ_{x_n},\tag{3.9}$$

where $i_1,\cdots,i_\lambda$ *take one of the values* $1,\cdots,n$, *while* ν_j *ranges over* λ_j *of the values* $1,\dots,\lambda$, *the number of factors in the product over* ν_j *being equal to* λ_j.

PROOF. Formula (3.9) is obviously true for $\lambda=0$. We prove its validity in the general case by induction.

Differentiating (3.9) with respect to $x_{i_{\lambda+1}}$, we get

$$\frac{\partial^{\lambda+1}Kf(x)}{\partial x_{i_1}\dots\partial x_{i_\lambda}\partial x_{i_{\lambda+1}}}$$

$$=\frac{(-1)^{\lambda+1}n\delta_{i_{\lambda+1}}^n}{cx_n^{n+1}}\sum_{\lambda_1+\dots+\lambda_n=\lambda}\int\prod_{j=1}^n\frac{\partial^{\lambda_j}}{\partial\xi_j^{\lambda_j}}\left[\prod_{\nu_j}^{\lambda_j}\left(\delta_j^{i_{\nu_j}}+\delta_{i_{\nu_j}}^n\frac{\xi_j-x_j}{x_n}\right)K^{(j)}\right]fdQ_{x_n}$$

$$+\frac{(-1)^\lambda}{cx_n^n}\sum_{\lambda_1+\dots+\lambda_n=\lambda}\sum_{k=1}^n\int\frac{\partial^{\lambda_k}}{\partial\xi_k^{\lambda_k}}\left\{\frac{\partial}{\partial x_{i_{\lambda+1}}}\left[\prod_{\nu_k}^{\lambda_k}\left(\delta_k^{i_{\nu_k}}+\delta_{i_{\nu_k}}^n\frac{\xi_k-x_k}{x_n}\right)K^{(k)}\right]\right\}$$

$$\times\prod_{\substack{j=1\\j\neq k}}\frac{\partial^{\lambda_j}}{\partial\xi_j^{\lambda_j}}\left[\prod_{\nu_j}^{\lambda_j}\left(\delta_j^{i_{\nu_j}}+\delta_{i_{\nu_j}}^n\frac{\xi_j-x_j}{x_n}\right)K^{(j)}\right]fdQ_{x_n}$$

$$=\frac{(-1)^{\lambda+1}}{cx_n^n}\sum_{\lambda_1+\dots+\lambda_n=\lambda}\sum_{k=1}^n\int\frac{\partial^{\lambda_k}}{\partial\xi_k^{\lambda_k}}\left\{\frac{\delta_{i_{\lambda+1}}^n}{x_n}\prod_{\nu_k}^{\lambda_k}\left(\delta_k^{i_{\nu_k}}+\delta_{i_{\nu_k}}^n\frac{\xi_k-x_k}{x_n}\right)K^{(k)}\right.$$

$$\left.-\frac{\partial}{\partial x_{i_{\lambda+1}}}\left[\prod_{\nu_k}^{\lambda_k}\left(\delta_k^{i_{\nu_k}}+\delta_{i_{\nu_k}}^n\frac{\xi_k-x_k}{x_n}\right)K^{(k)}\right]\right\}\times$$

[4] We have applied formula (3.6) here.

$$\times \prod_{\substack{j=1\\ j\neq k}}^{n} \frac{\partial^{\lambda_j}}{\partial \xi_j^{\lambda_j}} \Big[\prod_{\nu_j}^{\lambda_j} \Big(\delta_j^{i_{\nu_j}} + \delta_{i_{\nu_j}}^{n} \frac{\xi_j - x_j}{x_n} \Big) K^{(j)} \Big] f dQ_{x_n} \text{(5)}$$

$$= \frac{(-1)^{\lambda+1}}{c x_n^n} \sum_{\lambda_1+\ldots+\lambda_n=\lambda} \sum_{k=1}^{n} \int \frac{\partial^{\lambda_k+1}}{\partial \xi_k^{\lambda_k+1}} \Big[\prod_{\nu_k}^{\lambda_k} \Big(\delta_k^{i_{\nu_k}} + \delta_{i_{\nu_k}}^{n} \frac{\xi_k - x_k}{x_n} \Big) K^{(k)} \Big]$$

$$\times \prod_{\substack{j=1\\ j\neq k}}^{n} \frac{\partial^{\lambda_j}}{\partial \xi_j^{\lambda_j}} \Big[\prod_{\nu_j}^{\lambda_j} \Big(\delta_j^{i_{\nu_j}} + \delta_{i_{\nu_j}}^{n} \frac{\xi_j - x_j}{x_n} \Big) K^{(j)} \Big] f dQ_{x_n}$$

$$= \frac{(-1)^{\lambda+1}}{c x_n^n} \sum_{\lambda_1'+\ldots+\lambda_n'=\lambda+1} \int \prod_{j=1}^{n} \frac{\partial^{\lambda_j'}}{\partial \xi_j^{\lambda_j'}} \Big[\prod_{\nu_j}^{\lambda_j'} \Big(\delta_j^{i_{\nu_j}} + \delta_{i_{\nu_j}}^{n} \frac{\xi_j - x_j}{x_n} \Big) K^{(j)} f dQ_{x_n}.$$

[3.4]. *Suppose* $f \in L_p^{(n)}$ *and all of its weak derivatives up to order* μ *inclusively also belong to* $L_p^{(n)}$. *Then when* $x_n \neq 0$

$$\frac{\partial^{\lambda+\mu} Kf(x)}{\partial x_{i_1} \ldots \partial x_{i_{\lambda+\mu}}}$$

$$= \frac{(-1)^{\lambda}}{c x_n^n} \sum_{\lambda_1+\ldots+\lambda_n+\mu_1+\ldots+\mu_n=\lambda+\mu} \int \prod_{j=1}^{n} \frac{\partial^{\lambda_j}}{\partial \xi_j^{\lambda_j}} \Big[\prod_{\nu_j}^{\lambda_j+\mu_j} \Big(\delta_j^{i_{\nu_j}} + \delta_{i_{\nu_j}}^{n} \frac{\xi_j - x_j}{x_n} \Big) K^{(j)} \Big]$$

$$\times f_{\xi_1^{\mu_1} \ldots \xi_n^{\mu_n}} dQ_{x_n}, \tag{3.10}$$

and, in particular, when $i_1 = \cdots = i_{\lambda+\mu} = n$

$$\frac{\partial^{\lambda+\mu} Kf(x)}{\partial x_n^{\lambda+\mu}}$$

$$= \frac{(-1)^{\lambda}}{c x_n^n} \sum_{\lambda_1+\ldots+\lambda_n+\mu_1+\ldots+\mu_n=\lambda+\mu} \int \prod_{j=1}^{n} \frac{\partial^{\lambda_j}}{\partial \xi_j^{\lambda_j}} \Big[\Big(\delta_j^n + \frac{j - x_j}{x_n} \Big)^{\lambda_j+\mu_j} K^{(j)} \Big]$$

$$\times f_{\xi_1^{\mu_1} \ldots \xi_n^{\mu_n}} dQ_{x_n}. \tag{3.11}$$

These formulas are obtained by a direct μ multiple integration by parts of formula (3.9), which is possible here by virtue of the definition of weak derivatives

(5) We apply formula (3.8) here.

([2], page 33) and property (1.3) of the kernels $K^{(j)}$

Making the change of variable $\xi_i = x_i + x_n t_i/2$ in (3.10) and (3.11) and noting that $\partial/\partial\xi_i = (2/x_n)\partial/\partial t_i$, $i = 1, \cdots, n$, we get

$$\frac{\partial^{\lambda+\mu} Kf(x)}{\partial x_{i_1} \ldots \partial x_{i_{\lambda+\mu}}}$$

$$= \frac{(-1)^\lambda 2^{\lambda-n}}{c x_n^\lambda} \sum_{\lambda_1+\ldots+\lambda_n+\mu_1+\ldots+\mu_n=\lambda+\mu} \int_{-1}^{1} \cdots \int_{-1}^{1} \prod_{j=1}^{n} \frac{d^{\lambda_j}}{dt_j^{\lambda_j}} \Big[\prod_{\nu_j}^{\lambda_j+\mu_j} \Big(\delta_j^{i_{\nu_j}} + \frac{\delta^n_{i_{\nu_j}} t_j}{2}\Big) K(t_j) \Big]$$

$$\times f_{\xi_1^{\mu_1} \ldots \xi_n^{\mu_n}} \left(x_1 + \frac{x_n t_1}{2}, \ldots\right) dt_1 \ldots dt_n, \tag{3.12}$$

$$\frac{\partial^{\lambda+\mu} Kf(x)}{\partial x_n^{\lambda+\mu}}$$

$$= \frac{(-1)^\lambda 2^{\lambda-n}}{c x_n^\lambda} \sum_{\lambda_1+\ldots+\lambda_n+\mu_1+\ldots+\mu_n=\lambda+\mu} \int_{-1}^{1} \cdots \int_{-1}^{1} \prod_{j=1}^{n} \frac{d^{\lambda_j}}{dt_j^{\lambda_j}} \Big[\Big(\delta_j^n + \frac{t_j}{2}\Big)^{\lambda_j+\mu_j}$$

$$\times K(t_j)\Big] f_{\xi_1^{\mu_1} \ldots \xi_n^{\mu_n}} \left(x_1 + \frac{x_n t_1}{2}, \ldots\right) dt_1 \ldots dt_n. \tag{3.13}$$

We note that if one formally calculates the derivative $\partial^\mu Kf/\partial x_{i_1} \cdots \partial x_{i_\mu}$ by Leibniz' rule, starting from formula (1.9) for Kf, one obtains, as is easily verified by a direct calculation, precisely formula (3.12) for $\lambda = 0$. It follows from what has been proved by us that the differentiation of formula (1.9) by Leibniz' rule is valid within the limits of the existence of the weak derivatives of f.

[3.5]. *Let* $f \in L_p^{(n)}$ *and suppose the weak derivative*

$$\frac{\partial^\lambda f}{\partial x_1^{\lambda_1} \ldots \partial x_{n-1}^{\lambda_{n-1}}} \in L_p^{(n)}$$

exists. Then

$$\frac{\partial^\lambda Kf}{\partial x_1^{\lambda_1} \ldots \partial x_{n-1}^{\lambda_{n-1}}} = K \frac{\partial^\lambda f}{\partial x_1^{\lambda_1} \ldots \partial x_{n-1}^{\lambda_{n-1}}}, \quad \lambda_1 + \ldots + \lambda_{n-1} = \lambda. \tag{3.14}$$

For noting that among the functions $K^{(j)}$, $j = 1, \dots, n$, only $K^{(i)}$ depends on x_i, $i = 1, \dots, n-1$, applying formula (3.7) and integrating by parts, we get

$$\frac{\partial^\lambda Kf}{\partial x_1^{\lambda_1} \dots \partial x_{n-1}^{\lambda_{n-1}}} = \frac{1}{cx_n^n} \int K^{(n)} \prod_{j=1}^{n-1} \frac{\partial^{\lambda_j} K^{(j)}}{\partial x_j^{\lambda_j}} f dQ_{x_n}$$

$$= \frac{(-1)^\lambda}{cx_n^n} \int K^{(n)} \prod_{j=1}^{n-1} \frac{\partial^{\lambda_j} K^{(j)}}{\partial \xi_j^{\lambda_j}} f dQ_{x_n} = \frac{1}{cx_n^n} \int \prod_{j=1}^{n} K^{(j)} f_{\xi_1^{\lambda_1} \dots \xi_{n-1}^{\lambda_{n-1}}} dQ_{x_n}.$$

COROLLARY.

$$\left\| \frac{\partial^\lambda Kf}{\partial x_1^{\lambda_1} \dots \partial x_{n-1}^{\lambda_{n-1}}} \right\|_p^{(n)} \leqslant 2^{\frac{1}{p}} \left\| \frac{\partial^\lambda f}{\partial x_1^{\lambda_1} \dots \partial x_{n-1}^{\lambda_{n-1}}} \right\|_p^{(n)}, \quad \lambda_1 + \dots + \lambda_{n-1} = \lambda.$$

But for the derivatives involving differentiation with respect to x_n one can only assert, statting from formula (3.13) for $\lambda = 0$, that

$$\left\| \frac{\partial^\mu Kf}{\partial x_1^{\mu_1} \dots \partial x_n^{\mu_n}} \right\|_p^{(n)} \leqslant A \sum_{\nu_1 + \dots + \nu_n = \mu} \left\| \frac{\partial^\mu f}{\partial x_1^{\nu_1} \dots \partial x_n^{\nu_n}} \right\|_p^{(n)}$$

where $A > 0$ is a constant not depending on f.

For the sake of convenience we introduce the following terminology.

DEFINITION. A function $L(t)$ defined on the interval $-1 \leqslant t \leqslant 1$ is said to be a *kernel* if:

1°. $L(t)$ together with all of its derivatives is continuous on the interval $-1 \leqslant t \leqslant 1$;

2°. $L^{(s)}(-1) = L^{(s)}(1) = 0$ for $s = 0, 1, \dots$.

A kernel $L(t)$ is said to be a kernel *with the singular property* (*) if

$$\int_{-1}^{1} L(t)\, dt = 0. \qquad (*)$$

As examples of kernels we have the functions $K(t)$ and $L(t) = (a + bt)^\lambda K(t)$. An example of a singular kernel is $L(t) = d^s((a + bt)^\lambda K(t))/dt^s$, $s = 1, 2, \dots$.

[3.6]. *Suppose* $f \in H_{p(n)}^{(r)}(M)$, $r = \bar{r} + \alpha$. *Then every weak derivative* $\partial^{\lambda + \bar{r}} Kf / \partial x_{i_1} \cdots \partial x_{i_{\lambda+\bar{r}}}$ *can be represented by a linear combination of a finite number of terms each of which can be written in either of the following two forms:*

$$\frac{1}{x_n^\lambda}\int_{-1}^{1}\cdots\int_{-1}^{1}\prod_{j=1}^{n}L_j(t_j)\varphi\left(x_1+\frac{x_nt_1}{2},\ \ldots,\ x_n+\frac{x_nt_n}{2}\right)dt_1\ \ldots\ dt_n, \tag{3.15}$$

where

$$\varphi\in H_{p(n)}^{(\alpha)}(M'),\quad \|\varphi\|_p^{(n)}\leqslant M'; \tag{3.16}$$

or

$$\frac{1}{x_n^{\lambda+1}}\int_{-1}^{1}\cdots\int_{-1}^{1}\prod_{j=1}^{n}L_j(t_j)\psi\left(x_1+\frac{x_nt_1}{2},\ \ldots,\ x_n+\frac{x_nt_n}{2}\right)dt_1\ \ldots\ dt_n, \tag{3.17}$$

where

$$\psi\in H_{p(n)}^{(\alpha+1)}(M'),\quad \|\psi\|_p^{(n)}\leqslant M', \tag{3.18}$$

the constants $M'\leqslant c_1M+c_2\|f\|_p^{(n)}$, $c_1>0$ *and* $c_2>0$ *not depending on* M *or* f *and the functions* $L_j(t)$ *being kernels (generally different in* (3.15) *and* (3.17)) *among which there is at least one with the singular property* (*) *when* $\lambda>0$.

Formulas (3.15) and (3.17) follow directly from (3.12) and (3.13) and the fact that every weak derivative of order $\mu\leqslant\bar{r}$ of f belongs to the class $H_{p(n)}^{(r-\mu)}(M^{(\mu)})$, where $M^{(\mu)}\leqslant c_1^{(\mu)}M+c_2^{(\mu)}\|f\|_p^{(n)}$ (see [4], page 79, Theorem 3.2); while formula (3.17) follows from (3.15) by means of a single integration by parts (when $r>1$). Finally, from the fact that φ and ψ are weak derivatives of f, we get (3.16) and (3.18) according to a corresponding result of Nikol'skiĭ [4].

§4. The boundary values of Kf and its derivatives

A function f belonging to $L_p^{(n)}$ is defined in E^n to within a set of measure zero. Therefore, in order to be able to speak of its values on hyperplanes of lower dimension, we introduce the following definition (see [2, 4]): a function $\varphi(x_1,\ldots,x_m)$, $0<m<n$, is called the value of a function f *in the sense of convergence in the mean* (for fixed p, $1\leqslant p\leqslant\infty$) on an m-dimensional hyperplane $x_{m+1}=x_{m+1}^{(0)},\ldots,x_n=x_n^{(0)}$ if the function f can be so altered on a set of measure zero that the following limit exists:

$$\lim_{\sum_{i=m+1}^{n}h_i^2\to 0}\left\{\int_{-\infty}^{\infty}\cdots\int_{-\infty}^{\infty}|f(x_1,\ \ldots,\ x_m,\ x_{m+1}^{(0)}+h_{m+1},\ \ldots,\ x_n^{(0)}+h_n)\right.$$

$$\left.-\varphi(x_1,\ \ldots,\ x_m)|^p\,dx_1\ \ldots\ dx_m\right\}^{\frac{1}{p}}=0. \tag{4.1}$$

The boundary value defined in this way of a function f is unique to within a class of equivalent functions relative to an m-dimensional measure. We will denote it by

$$f\,|_{x_{m+1}=x^{(0)}_{m+1},\ \dots,\ x_n=x^{(0)}_n}$$

or simply by $f(x_1, \dots, x_m, x^{(0)}_{m+1}, \dots, x^{(0)}_n)$.

For a function $f \in L^{(n)}_p$ for which the value

$$f(x_1, \dots, x_m, x^{(0)}_{m+1}, \dots, x^{(0)}_n),$$

exists we introduce the notation

$$\| f \|^{(m)}_p = \left\{ \int_{-\infty}^{\infty} \cdots \int_{-\infty}^{\infty} | f(x_1, \dots, x_m, x^{(0)}_{m+1}, \dots, x^{(0)}_n) |^p dx_1 \dots dx_m \right\}^{\frac{1}{p}},$$

where $x^{(0)}_{m+1}, \dots, x^{(0)}_n$ are fixed.

The order of convergence to zero in equality (4.1) was found by Nikol'skiĭ. This result, which is formulated more precisely below, was announced by Nikol'skiĭ at a seminar on the theory of functions at the Mathematical Institute of the Academy of Sciences of the USSR in 1952 but was never published. Therefore Nikol'skiĭ's theorem—actually a somewhat modified form of it that is more suitable for our purposes—is given here together with a proof.(6)

[4.1] (Nikol'skiĭ's Theorem). *Suppose* $f \in H^{(r_1, \dots, r_n)}_{p(n)}(M)$, $1 \leqslant m < n$,

$$\beta_j = r_j \left(1 - \frac{1}{p} \sum_{m+1}^{n} \frac{1}{r_i}\right),$$

with $0 < \beta_j \leqslant 1, j = m+1, \dots, n$, *and suppose the quantities* $x_{m+1}, \dots, x_n$ *and* $h_{m+1}, \dots, h_n$ *have been arbitrarily fixed. Then there exist constants* $c_1 > 0$ *and* $c_2 > 0$ *not depending on* M, f *or* $x_{m+1}, \dots, x_n, h_{m+1}, \dots, h_n$ *such that*

(6) This is done with the amiable consent of S. M. Nikol'skiĭ.

$$\|\Delta^{(1)}_{x_{m+1}\dots x_n}(f;\ h_{m+1},\ \dots,\ h_n)\|^{(m)}_p$$

$$=\Big\{\int_{-\infty}^{\infty}\dots\int_{-\infty}^{\infty}|\,f(x_1,\ \dots,\ x_m,\ x_{m+1},\ \dots,\ x_n)$$

$$-f(x_1,\ \dots,x_m,\ x_{m+1}+h_{m+1},\ \dots,\ x_n+h_n)\,|^p dx_1\dots dx_m\Big\}^{\frac{1}{p}}$$

$$\leqslant(c_1M+c_2\|f\|^{(n)}_p)\Big(\sum_{\beta_j<1}|h_j|^{\beta_j}+\sum_{\beta_j=1}|h_j|\ln|h_j|\Big). \tag{4.2}$$

COROLLARY. *Under the assumptions of the theorem, for any* $\epsilon>0$

$$\|\Delta^{(1)}_{x_{m+1},\ \dots,\ x_n}(f;\ h_{m+1},\ \dots,\ h_n)\|^{(n)}_p$$

$$\leqslant(c_1'M+c_2'\|f\|^{(n)}_p)\Big(\sum_{\beta_j<1}|h_j|^{\beta_j}+\frac{1}{\varepsilon}\sum_{\beta_j=1}|h_j|^{1-\varepsilon}\Big). \tag{4.3}$$

PROOF. As we already know, the fact that $f\in H^{(r_1,\dots,r_n)}_{p(n)}(M)$ implies (see (2.15)–(2.17)) that in $L^{(n)}_p$ we can expand it in entire functions q_s of exponential type of degrees $2^{s/r_i}$ respectively in the variables x_i, $i=1,\dots,n$:

$$f=\sum_{s=0}^{\infty}q_s, \tag{4.4}$$

where

$$\|q_s\|^{(n)}_p<\frac{c}{2^s},\quad c\leqslant a_1M_i+a_2\|f\|^{(n)}_p,\quad s=0,\ 1,\ \dots. \tag{4.5}$$

Therefore

$$\|\Delta^{(1)}_{x_{m+1},\dots,x_n}(f,\ h_{m+1},\ \dots,\ h_n)\|^{(m)}_p$$

$$\leqslant\sum_{s=0}^{\infty}\|q_s(x_1,\ \dots,\ x_m,\ x_{m+1}+h_{m+1},\ \dots,\ x_n+h_n)$$

$$-q_s(x_1,\ \dots,\ x_m,\ x_{m+1}\ \dots\ x_n)\|^{(m)}_p \tag{4.6}$$

$$\leqslant\sum_{s=0}^{\infty}\sum_{j=m+1}^{n}\|q_s(x_1,\ \dots,\ x_m,\ \dots,\ x_{j-1},\ x_j+h_j,\ x_{j+1}+h_{j+1},\ \dots,x_n+h_n)$$

$$-q_s(x_1,\ \dots,\ x_m,\ \dots,\ x_j,\ x_{j+1}+h_{j+1},\ \dots,\ x_n+h_n)\|^{(m)}_p.$$

We now note that for any entire function $g_{\nu_1\ldots\nu_n}(x_1, \cdots, x_n)$ of exponential type of degrees ν_i respectively in the variables x_i, $i = 1, \ldots, n$, and for any fixed $x_{m+1}, \ldots, x_n$ we have the inequality ([3], page 11)

$$\| g_{\nu_1\ldots\nu_n} \|_p^{(m)} \leqslant 2^n \left(\prod_{j=m+1}^{n} \nu_j \right)^{\frac{1}{p}} \| g_{\nu_1\ldots\nu_n} \|_p^{(n)}. \tag{4.7}$$

The function

$$q_s(x_1, \ldots, x_{j-1}, x_j + h_j, \cdots, x_n + h_n)$$

$$-q_s(x_1, \ldots, x_j, x_{j+1}+h_{j+1}, \ldots, x_n+h_n)$$

is an entire function of degree $2^{s/r_i}$ in the variable x_i, $i = 1, \ldots, n$, for fixed $h_{m+1}, \ldots, h_n$. Therefore, applying (4.7) to (4.6) and making a change of variable, we get

$$\| \Delta^{(1)}_{x_{m+1}\ldots x_n}(f;\ h_{m+1}, \ldots, h_n) \|_p^{(m)}$$

$$\leqslant 2^n \sum_{j=m+1}^{n} \sum_{s=0}^{\infty} 2^{\frac{s}{p} \sum_{i=m+1}^{n} \frac{1}{r_i}} \| q_s(x_1, \ldots, x_{j-1}, x_j + h_j, \ldots, x_n + h_n)$$

$$- q_s(x_1, \ldots, x_j, x_{j+1} + h_{j+1}, \ldots, x_n + h_n) \|_p^{(n)}$$

$$\leqslant 2^n \sum_{j=m+1}^{n} \sum_{s=0}^{\infty} 2^{\frac{s}{p} \sum_{i=m+1}^{n} \frac{1}{r_i}} \| q_s(x_1, \ldots, x_{j-1}, x_j + h_j, x_{j+1}, \ldots, x_n)$$

$$- q_s(x_1, \ldots, x_{j-1}, x_j, x_{j+1}, \ldots, x_n) \|_p^{(n)}$$

$$= 2^n \sum_{j=m+1}^{n} \sum_{s=0}^{\infty} 2^{s\left(1 - \frac{\beta_j}{r_j}\right)} \| \Delta^{(1)}_{x_j}(q_s, h_j) \|_p^{(n)}. \tag{4.8}$$

Let us estimate the inner sum. To this end we fix a value of j, choose an integer $N_j \geqslant 0$ such that

$$\frac{1}{2^{N_j+1}} \leqslant | h_j |^{r_j} < \frac{1}{2^{N_j}} \tag{4.9}$$

(the estimate is trivial in the case $|h_j| \geqslant 1$) and split the sum being considered into two sums:

$$\sum_{s=0}^{\infty} 2^{s\left(1-\frac{\beta_j}{r_j}\right)} \| \Delta_{x_j}^{(1)}(q_s, h_j) \|_p^{(n)} = \sum_0^{N_j} + \sum_{N_j+1}^{\infty}.$$

Applying Lemma 3 of the paper [3] (page 15) of Nikol′skiĭ, the generalized Bernstein inequality (loc. cit., page 4) and inequality (4.5), we get

$$\sum_{s=0}^{N_j} 2^{s\left(1-\frac{\beta_j}{r_j}\right)} \| \Delta_{x_n}^{(1)}(q_s, h_j) \|_p^{(n)} \leqslant \sum_{s=0}^{N_j} 2^{s\left(1-\frac{\beta_j}{r_j}\right)} \left\| \frac{\partial q_s}{\partial x_j} \right\|_p^{(n)} | h_j |$$

$$\leqslant | h_j | \sum_{s=0}^{N_j} 2^{s\left(1-\frac{\beta_j}{r_j}\right)} 2^{\frac{s}{r_j}} \| q_s \|_p^{(n)} \leqslant c | h_j | \sum_{s=0}^{N_j} 2^{\frac{s}{r_j}(1-\beta_j)}$$

$$= \begin{cases} c | h_j | N_j \leqslant \gamma_1^{(j)} c | h_j | \ln \dfrac{1}{|h_j|} & \text{for } \beta_j = 1, \\ c | h_j | \dfrac{2^{\frac{(1-\beta_j)(N_j+1)}{r_j}} - 1}{2^{\frac{1-\beta_j}{r_j}} - 1} \leqslant \gamma_2^{(j)} c | h_j | 2^{N_j \frac{1-\beta_j}{r_j}} < \gamma_2^{(j)} c | h_j |^{\beta_j} & \text{for } \beta_j < 1. \end{cases}$$

Further,

$$\sum_{s=N_j+1}^{\infty} 2^{s\left(1-\frac{\beta_j}{r_j}\right)} \| \Delta_{x_j}^{(1)}(q_s, h_j) \|_p^{(n)} \leqslant 2 \sum_{s=N_j+1}^{\infty} 2^{s\left(1-\frac{\beta_j}{r_j}\right)} \| q_s \|_p^{(n)}$$

$$\leqslant 2c \sum_{s=N_j+1}^{\infty} 2^{-\frac{s\beta_j}{r_j}} \leqslant \gamma_3^{(j)} c 2^{\frac{\beta_j(N_j+1)}{r_j}} \leqslant \gamma_3^{(j)} c | h |^{\beta_j},$$

where $\gamma_1^{(j)}, \gamma_2^{(j)}, \gamma_3^{(j)}$ are certain constants depending only on r_j and β_j. Combining these inequalities and setting $\gamma = \max_j \{\gamma_1^{(j)}, \gamma_2^{(j)}, \gamma_3^{(j)}\}$, we obtain the estimate

$$\sum_{s=0}^{\infty} 2^{s\left(1-\frac{\beta_j}{r_j}\right)} \| \Delta_{x_j}^{(1)}(q_s, h_j) \|_p^{(n)} \leqslant \begin{cases} \gamma c | h_j | \ln \dfrac{1}{|h_j|} & \text{for } \beta_j = 1, \\ \gamma c | h_j |^{\beta_j} & \text{for } \beta_j < 1, \end{cases}$$

$$j = m+1, \ldots, n.$$

Returning to inequality (4.8), we get

$$\| \Delta^{(1)}_{x_{m+1},\dots,x_n}(f;\ h_{m+1},\ \dots,\ h_n) \|^{(m)}_{p}$$
$$\leqslant 2^n \gamma c \Big(\sum_{\beta_j<1} |h_j|^{\beta_j} + \sum_{\beta_j=1} |h_j| \ln \frac{1}{|h_j|} \Big),$$

and since the constant c satisfies an inequality of form (4.5) while γ depends only on r_j and β_j, $j = m+1, \dots, n$, Theorem [4.1] is proved.

To prove the corollary of Theorem [4.1], which we essentially use in the sequel, it suffices to note that $0 < h^{\epsilon} \ln h^{-1} \leqslant \epsilon^{-1}$ for $0 < h \leqslant 1$.

We also note that estimates of a similar kind, characterizing the rate of convergence of a function to its boundary values, were obtained earlier by Sobolev for the functions of L_2 and by Kondrašov for the functions of an arbitrary space L_p [8].

Theorem [4.1] receives a more finished formulation, which does not depend on whether $\beta_j < 1$ or $\beta_j = 1$ and is precisely the one proved by Nikol'skiĭ, if one introduces the following definition.

A function $\varphi(x_1, \dots, x_m)$ is called the *value* of a function $f(x_1, \dots, x_n)$ in the mean for fixed values of $x_{m+1}, \dots, x_n$ (and for given p, $1 \leqslant p < \infty$) if

$$\lim_{\sum_{j=m+1}^{n} h_j^2 \to 0} \Big\{ \int_{-\infty}^{\infty} \cdots \int_{-\infty}^{\infty} | f(x_1,\ \dots,\ x_m,\ x_{m+1} - h_{m+1},\ \dots,\ x_n - h_n)$$
$$- 2\varphi(x_1,\ \dots,\ x_m) \tag{4.10}$$
$$+ f(x_1,\ \dots,\ x_m, x_{m+1} + h_{m+1},\ \dots,\ x_n + h_n) |^p dx_1 \dots dx_m \Big\}^{\frac{1}{p}} = 0.$$

It is obvious that condition (4.1) implies condition (4.10). We now put

$$\Delta^{(2)}_{x_{m+1},\dots,x_n}(f;\ h_{m+1},\ \dots,\ h_n)$$
$$= f(x_1,\ \dots,\ x_m,\ x_{m+1} - h_{m+1},\ \dots,\ x_n - h_n) - 2f(x_1,\ \dots,\ x_n)$$
$$+ f(x_1,\ \dots,\ x_m,\ x_{m+1} + h_{m+1},\ \dots,\ x_n + h_n),$$

and assert that under the assumption

$$0 < \beta_j = r_j\left(1 - \frac{1}{p}\sum_{i=m+1}^{n}\frac{1}{r_i}\right) \leqslant 1$$

we will have for every function $f \in H_p^{(r_1,\ldots,r_n)}(M)$

$$\| \Delta^{(2)}_{x_{m+1},\ldots,x_n}(f;\ h_{m+1},\ldots,\ h_n) \|_p^{(m)}$$

$$\leqslant (c_1 M + c_2 \| f \|_p^{(n)}) \sum_{j=m+1}^{n} | h_j |^{\beta_j}.$$

Let us prove this assertion. Expanding f in the appropriate entire functions q_s (see (4.4)) and arguing analogously to the proof of Theorem [4.1], we get

$$\| \Delta^{(2)}_{x_{m+1}\cdots x_n}(f;\ h_{m+1},\ \ldots,\ h_n) \|_p^{(m)}$$

$$\leqslant 2^n \sum_{j=m+1}^{n}\sum_{s=0}^{\infty} 2^{s\left(1-\frac{\beta_j}{r_j}\right)} \| \Delta^{(2)}_{x_j}(q_s,\ h_j) \|_p^{(n)}.$$

For each j we choose $N_j > 0$ according to inequality (4.9). Then

$$\sum_{s=0}^{\infty} 2^{s\left(1-\frac{\beta_j}{r_j}\right)} \| \Delta^{(2)}_{x_j}(q_s,\ h_j) \|_p^{(n)}$$

$$\leqslant \sum_{s=0}^{N_j} 2^{s\left(1-\frac{\beta_j}{r_j}\right)} \| \Delta^{(2)}_{x_j}(q_s,\ h_j) \|_p^{(n)} + 4 \sum_{s=N_j+1}^{\infty} 2^{s\left(1-\frac{\beta_j}{r_j}\right)} \| q_s \|_p^{(n)}.$$

But

$$\sum_{s=0}^{N_j} 2^{s\left(1-\frac{\beta_j}{r_j}\right)} \| \Delta^{(2)}_{x_j}(q_s,\ h_j) \|_p^{(n)} \leqslant | h_j |^2 \sum_{s=0}^{N_j} 2^{s\left(1-\frac{\beta_j}{r_j}\right)} \left\| \frac{\partial^2 q_s}{\partial x_j^2} \right\|_p^{(n)}$$

$$\leqslant | h_j |^2 \sum_{s=0}^{N_j} 2^{s\left(1-\frac{\beta_j}{r_j}\right)} 2^{\frac{2s}{r_j}} \| q_s \|_p^{(n)} \leqslant c\, | h_j |^2 \sum_{s=0}^{N_j} 2^{\frac{s}{r_j}(2-\beta_j)}$$

$$= c\, | h_j |^2 \frac{2^{\frac{(2-\beta_j)(N_j+1)}{r_j}} - 1}{2^{\frac{2-\beta_j}{r_j}} - 1} \leqslant \gamma'^{(j)} c\, | h_j |^2 2^{\frac{N_j(2-\beta_j)}{r_j}} \leqslant \gamma'^{(j)} c\, | h_j |^{\beta_j};$$

$$4\sum_{s=N_j+1}^{\infty} 2^{s\left(1-\frac{\beta_j}{r_j}\right)} \| q_s \|_p^{(n)} \leqslant 4c \sum_{s=N_j+1}^{\infty} 2^{-\frac{s\beta_j}{r_j}} \leqslant \gamma''^{(j)} c 2^{-\frac{\beta_j(N_j+1)}{r_j}}$$

$$\leqslant \gamma''^{(j)} c \, | h_j |^{\beta_j}.$$

If now $\beta = \max_j \{\gamma'^{(j)}; \gamma''^{(j)}\}$, then

$$\| \Delta^{(2)}_{x_{m+1},\ldots,x_n} (f; h_{m+1}, \ldots, h_n) \|_p^{(m)} \leqslant 2^n c\gamma \sum_{j=m+1}^{n} | h_j |^{\beta_j},$$

where c satisfies an inequality of form (4.5), and the assertion follows.

For the sake of simplicity we will adhere to definition (4.1) in the sequel.

[4.2]. *Suppose* $\varphi \in L_p^{(n)}$ *and has the boundary value* $\varphi(x_1, \ldots, x_{n-1}, 0) \in L_p^{(n-1)}$:

$$\lim_{h\to 0} \int_{-\infty}^{\infty} \cdots \int_{-\infty}^{\infty} | \varphi(x_1, \ldots, x_{n-1}, h) - \varphi(x_1, \ldots, x_{n-1}, 0) |^p \, dx_1 \ldots dx_{n-1} = 0. \tag{4.11}$$

Suppose, further,

$$L\varphi(x_1, \ldots, x_n) = \int_{-1}^{1} \cdots \int_{-1}^{1} \prod_{j=1}^{n} L_j(t_j) \varphi\left(x_1 + \frac{x_n t_1}{2}, \ldots, x_n + \frac{x_n t_n}{2}\right) dt_1 \ldots dt_n,$$

where the $L_j(t)$ *are kernels. Then* $L\varphi$ *has the boundary value*

$$L\varphi(x_1, \ldots, x_{n-1}, 0) = \varphi(x_1, \ldots, x_{n-1}, 0) \int_{-1}^{1} \cdots \int_{-1}^{1} \prod_{j=1}^{n} L_j(t_j) \, dt_1 \ldots dt_n. \tag{4.12}$$

PROOF. We estimate the norm:

$$\left\{ \int_{-\infty}^{\infty} \cdots \int_{-\infty}^{\infty} \left| L\varphi(x_1, \ldots, x_n) - \varphi(x_1, \ldots, x_{n-1}, 0) \right.\right.$$

$$\times \left.\left. \int_{-1}^{1} \cdots \int_{-1}^{1} \prod_{j=1}^{n} L_j(t_j)\, dt_1 \ldots dt_n \right|^p dx_1 \ldots dx_{n-1} \right\}^{\frac{1}{p}}$$

$$= \left\{ \int_{-\infty}^{\infty} \cdots \int_{-\infty}^{\infty} \left| \int_{-1}^{1} \cdots \int_{-1}^{1} \prod_{j=1}^{n} L_j(t_j) \left[\varphi\left(x_1 + \frac{x_n t_1}{2}, \ldots, x_n + \frac{x_n t_n}{2}\right) \right.\right.\right.$$

$$\left.\left.\left. - \varphi(x_1 \ldots x_{n-1}, 0) \right] dt_1 \ldots dt_n \right|^p dx_1 \ldots dx_{n-1} \right\}^{\frac{1}{p}}$$

$$\leqslant \int_{-1}^{1} \cdots \int_{-1}^{1} \prod_{j=1}^{n} |L_j(t_j)| \left\{ \int_{-\infty}^{\infty} \cdots \int_{-\infty}^{\infty} \left| \varphi\left(x_1 + \frac{x_n t_1}{2}, \ldots, x_n + \frac{x_n t_n}{2}\right) \right.\right.$$

$$\left.\left. - \varphi(x_1, \ldots, x_{n-1}, 0) \right|^p dx_1 \ldots dx_{n-1} \right\}^{\frac{1}{p}} dt_1 \ldots dt_n{}^{(7)}$$

$$\leqslant \int_{-1}^{1} \cdots \int_{-1}^{1} \prod_{j=1}^{n} |L_j(t_j)| \left\{ \int_{-\infty}^{\infty} \cdots \int_{-\infty}^{\infty} \left| \varphi\left(x_1 + \frac{x_n t_1}{2}, \ldots, x_n + \frac{x_n t_n}{2}\right) \right.\right.$$

$$\left.\left. - \varphi\left(x_1 + \frac{x_n t_1}{2}, \ldots, x_{n-1} + \frac{x_n t_{n-1}}{2}, 0\right) \right|^p dx_1 \ldots dx_{n-1} \right\}^{\frac{1}{p}} dt_1 \ldots dt_n$$

$$+ \int_{-1}^{1} \cdots \int_{-1}^{1} \prod_{j=1}^{n} |L_j(t_j)| \left\{ \int_{-\infty}^{\infty} \cdots \int_{-\infty}^{\infty} \left| \varphi\left(x_1 + \frac{x_n t_1}{2}, \ldots, x_{n-1} + \frac{x_n t_{n-1}}{2}, 0\right) \right.\right.$$

$$\left.\left. - \varphi(x_1, \ldots, x_{n-1}, 0) \right|^p dx_1 \ldots dx_{n-1} \right\}^{\frac{1}{p}} dt_1 \ldots dt_n = I_1 + I_2.$$

In the integral I_1 we make the change of variable $u_i = x_i + x_n t_i/2$, $i = 1, \ldots, n-1$. We then have

$$I_1 = \int_{-1}^{1} \cdots \int_{-1}^{1} \prod_{j=1}^{n} |L_j(t_j)| \left\{ \int_{-\infty}^{\infty} \cdots \int_{-\infty}^{\infty} \left| \varphi\left(u_1, \ldots, u_{n-1}, x_n + \frac{x_n t_n}{2}\right) \right.\right.$$

$$\left.\left. - \varphi(u_1, \ldots, u_{n-1}, 0) \right|^p du_1 \ldots du_{n-1} \right\}^{\frac{1}{p}} dt_1 \ldots dt_n.$$

(7) We have applied the generalized Minkowski inequality here.

But by virtue of (4.11) the inner integral of this expression tends to zero as $x_n \to 0$ (uniformly in t_n), and hence $\lim_{x_n \to 0} I_1 = 0$. Further, by virtue of the hypothesis of the theorem the function φ on the hyperplane $x_n = 0$ belongs to the space $L_p^{(n-1)}$ and is therefore continuous in the large relative to this space (see [2], page 11), and [28], English page 162, i.e.

$$\left\{ \int_{-\infty}^{\infty} \cdots \int_{-\infty}^{\infty} \left| \varphi \left(x_1 + \frac{x_n t_1}{2}, \ldots, x_{n-1} + \frac{x_n t_{n-1}}{2}, 0 \right) - \varphi(x_1, \ldots, x_{n-1}, 0) \right|^p dx_1 \ldots dx_{n-1} \right\}^{\frac{1}{p}} \to 0 \quad \text{for } x_n \to 0$$

(uniformly in $t_1, \ldots, t_{n-1}$), and hence $\lim_{x_n \to 0} I_2 = 0$, which proves (4.12).

[4.3]. *Suppose* $f \in L_p^{(n)}$ *and there exists the weak derivative*

$$\frac{\partial^\lambda f}{\partial x_1^{\lambda_1} \ldots \partial x_{n-1}^{\lambda_{n-1}}} \in H_{p(n)}^{(r)}, \quad r > \frac{1}{p}, \quad \lambda_1 + \ldots + \lambda_{n-1} = \lambda.$$

Then

$$\frac{\partial^\lambda Kf(x_1, \ldots, x_{n-1}, 0)}{\partial x_1^{\lambda_1} \ldots \partial x_{n-1}^{\lambda_{n-1}}} = \frac{\partial^\lambda f(x_1, \ldots, x_{n-1}, 0)}{\partial x_1^{\lambda_1} \ldots \partial x_{n-1}^{\lambda_{n-1}}}. \tag{4.13}$$

In particular, when $\lambda = 0$

$$Kf(x_1, \ldots, x_{n-1}, 0) = f(x_1, \ldots, x_{n-1}, 0). \tag{4.14}$$

These formulas are a direct consequence of [4.2], (3.14) and (1.15). The situation is somewhat more complicated with derivatives of the type $\partial^\lambda Kf/\partial x_n^\lambda$.

[4.4]. *Suppose* $f \in L_p^{(n)}$ *and all of its weak derivatives up to order* μ *inclusively exist, each derivative of order* μ *belonging to the class* $H_p^{(r)}$, *where* $r > 1/p$. *Then*

$$\frac{\partial^\mu Kf(x_1, \ldots, x_{n-1}, 0)}{\partial x_n^\mu} = \sum_{\mu_1 + \ldots + \mu_n = \mu} c_{\mu_1 \ldots \mu_n} \frac{\partial^\mu f(x_1, \ldots, x_{n-1}, 0)}{\partial x_1^{\mu_1} \ldots \partial x_n^{\mu_n}}, \tag{4.15}$$

$$c_{0 \ldots 0 \mu} \neq 0. \tag{4.16}$$

This follows from [4.2] and formula (3.13) for $\lambda = 0$. In this connection,

$$c_{\mu_1 \ldots \mu_n} = \frac{1}{2^n c} \int_{-1}^{1} \cdots \int_{-1}^{1} \prod_{j=1}^{n} \left(\delta_j^n + \frac{t_j}{2} \right)^{\mu_j} K(t_j)\, dt_1 \ldots dt_n;$$

in particular,

$$c_{0 \ldots 0\mu} = \frac{1}{2^n c} \int_{-1}^{1} \cdots \int_{-1}^{1} \prod_{j=1}^{n} K(t_j) \left(1 + \frac{t_n}{2} \right)^{\mu} dt_1 \ldots dt_n$$

$$> \frac{1}{2^{n+\mu} c} \int_{-1}^{1} \cdots \int_{-1}^{1} \prod_{j=1}^{n} K(t_j)\, dt_1 \ldots dt_n = \frac{1}{2^{\mu}}.$$

§5. Function extensions from an $(n-1)$-dimensional hyperplane onto the whole space

Let $\sigma(x_1, \ldots, x_n)$ be a nonnegative function defined on E^n and, for any function f defined on E^n, let

$$\| f \|_{p(\sigma,\nu)}^{(n)} = \left\{ \int_{-\infty}^{\infty} \cdots \int_{-\infty}^{\infty} \sigma^{\nu}(x_1, \ldots, x_n) \, | f(x_1, \ldots, x_n) |^p dx_1 \ldots dx_n \right\}^{\frac{1}{p}}. \tag{5.1}$$

We will call $\|f\|_{p(\sigma,\nu)}^{(n)}$ the norm of f with weight σ^{ν} in the sense of $L_p^{(n)}$.

Suppose now $f \in L_p^{(n)}$, the $L_j(t)$ $(j = 1, \ldots, n)$ are kernels and

$$Lf = \int_{-1}^{1} \cdots \int_{-1}^{1} \prod_{j=1}^{n} L_j(t_j) f\left(x_1 + \frac{x_n t_1}{2}, \ldots, x_n + \frac{x_n t_n}{2} \right) dt_1 \ldots dt_n.$$

Then

$$\| Lf \|_p^{(n)} \leqslant A \| f \|_p^{(n)}, \tag{5.2}$$

where A is a constant. In particular, $Lf \in L_p^{(n)}$ if $f \in L_p^{(n)}$. The proof of this is completely analogous to that of [1.1].

[5.1]. *Supppse* $f \in H_{p(n)}^{(r)}(M)$, $r = \bar{r} + \alpha$, $r > 1/p$ *and* $1 \leqslant p < \infty$. *Then for any* $\epsilon > 0$ *and* $s = 1, 2, \cdots$,

$$\left\| \frac{\partial^{s+\bar{r}} Kf}{\partial x_1^{s_1} \ldots \partial x_n^{s_n}} \right\|_{p[\,|x_n|, p(s-\alpha)+\varepsilon]}^{(n)} \leqslant \frac{a_1^{(s)} M + a_2^{(s)} \| f \|_p^{(n)}}{\varepsilon^{1+\frac{1}{p}}}, \tag{5.3}$$

where the constants $a_1^{(s)}$ *and* $a_2^{(s)}$ *do not depend on* M, f *or* ϵ.

PROOF. First suppose $\alpha > 1/p$. Then by virtue of [3.6] the derivative $\partial^{s+\bar{r}} Kf/\partial x_1^{s_1} \cdots \partial x_n^{s_n}$ is representable by a linear combination of a finite number of terms of the form

$$\frac{1}{x_n^s} \int_{-1}^{1} \cdots \int_{-1}^{1} \prod_{j=1}^{n} L_j(t_j) \varphi \left(x_1 + \frac{x_n t_1}{2}, \ldots, x_n + \frac{x_n t_n}{2} \right) dt_1 \ldots dt_n, \tag{5.4}$$

$$\varphi \in H_p^{(\alpha)}(M'),$$

where

$$\| \varphi \|_p^{(n)} \leqslant M', \quad M' \leqslant c_1 M + c_2 \| f \|_p^{(n)} \tag{5.5}$$

and at least one of the kernels L_j is a kernel with the singular property (*).(8) It therefore suffices to obtain an estimate of type (5.3) for the expression (5.4). We have

$$I = \Bigg\{ \int_{-\infty}^{\infty} \cdots \int_{-\infty}^{\infty} | x_n |^{p(s-\alpha)+\varepsilon} \Bigg| \frac{1}{x_n^s} \int_{-1}^{1} \cdots \int_{-1}^{1} \prod_{j=1}^{n} L_j(t_j)$$

$$\times \varphi \left(x_1 + \frac{x_n t_1}{2}, \ldots, x_n + \frac{x_n t_n}{2} \right)$$

$$\times dt_1 \ldots dt_n \Bigg|^p dx_1 \ldots dx_n \Bigg\}^{\frac{1}{p}} \leqslant I_1 + I_2 + I_3,$$

$$I_1 = \Bigg\{ \int_{-\infty}^{\infty} \cdots \int_{-\infty}^{\infty} dx_1 \ldots dx_{n-1} \int_{1}^{\infty} \frac{1}{| x_n |^{p\alpha - \varepsilon}} \Bigg| \int_{-1}^{1} \cdots \int_{-1}^{1} \prod_{j=1}^{n} L_j(t_j)$$

$$\times \varphi \left(x_1 + \frac{x_n t_1}{2}, \ldots, x_n + \frac{x_n t_n}{2} \right) dt_1 \ldots dt_n \Bigg|^p dx_n \Bigg\}^{\frac{1}{p}},$$

$$I_2 = \Bigg\{ \int_{-\infty}^{\infty} \cdots \int_{-\infty}^{\infty} dx_1 \ldots dx_{n-1} \int_{-\infty}^{-1} \frac{1}{|x_n|^{p\alpha - \varepsilon}} \Bigg| \int_{-1}^{1} \cdots \int_{-1}^{1} \prod_{j=1}^{n} L_j(t_j)$$

$$\times \varphi \left(x_1 + \frac{x_n t_1}{2}, \ldots, x_n + \frac{x_n t_n}{2} \right) dt_1 \ldots dt_n \Bigg|^p dx_n \Bigg\}^{\frac{1}{p}},$$

(8) See page 28.

$$I_3=\Big\{\int\limits_{-\infty}^{\infty}\cdots\int\limits_{-\infty}^{\infty}dx_1\ldots dx_{n-1}\int\limits_{-1}^{1}\frac{1}{|x_n|^{p\alpha-\varepsilon}}\Big|\int\limits_{-1}^{1}\cdots\int\limits_{-1}^{1}\prod_{j=1}^{n}L_j(t_j)$$

$$\times\varphi\Big(x_1+\frac{x_nt_1}{2},\ldots,x_n+\frac{x_nt_n}{2}\Big)dt_1\ldots dt_n\Big|^p dx_n\Big\}^{\frac{1}{p}}.$$

The integrals I_1 and I_2 are estimated in the same way. For I_1, for example, we get, assuming $p\alpha>\epsilon$,

$$I_1=\Big\{\int\limits_{-\infty}^{\infty}\cdots\int\limits_{-\infty}^{\infty}\int\limits_{1}^{\infty}\frac{1}{|x_n|^{p\alpha-\varepsilon}}\Big|\int\limits_{1}^{1}\cdots\int\limits_{1}^{1}\prod_{j\ 1}^{n}L_j(t_j)$$

$$\times\varphi\Big(x_1+\frac{x_nt_1}{2},\ldots,x_n+\frac{x_nt_n}{2}\Big)dt_1\ldots dt_n\Big|^p dx_1\ldots dx_n\Big\}^{\frac{1}{p}}$$

$$\leqslant\Big\{\int\limits_{-\infty}^{\infty}\cdots\int\limits_{-\infty}^{\infty}\int\limits_{1}^{\infty}\Big|\int\limits_{-1}^{1}\cdots\int\limits_{1}^{1}\prod_{j=1}^{n}L_j(t_j)$$

$$\times\varphi\Big(x_1+\frac{x_nt_1}{2},\ldots,x_n+\frac{x_nt_n}{2}\Big)dt_1\ldots dt_n\Big|^p dx_1\ldots dx_n\Big\}^{\frac{1}{p}}\leqslant\|L\varphi\|_p^{(n)},$$

whence by virtue of (5.2)

$$I_1\leqslant A\,\|\varphi\|_p^{(n)}.$$

Applying (5.5), we conclude that

$$I_1\leqslant a_1^{\prime(s)}M+a_2^{\prime(s)}\|f\|_p^{(n)},\quad I_2\leqslant a_1^{\prime(s)}M+a_2^{\prime(s)}\|f\|_p^{(n)}.\tag{5.6}$$

We divide the estimation of I_3 into two cases. In the first case we assume that among the kernels $L_1,\ldots,L_{n-1}$ there is one with the singular property (*), for example, L_1. Then

$$I_3=\Big\{\int\limits_{-\infty}^{\infty}\cdots\int\limits_{-\infty}^{\infty}\int\limits_{-1}^{1}\frac{1}{|x_n|^{p\alpha-\varepsilon}}\Big|\int\limits_{-1}^{1}\cdots\int\limits_{-1}^{1}\prod_{j=1}^{n}L_j(t_j)$$

$$\times\varphi\Big(x_1+\frac{x_nt_1}{2},\ldots,x_n+\frac{x_nt_n}{2}\Big)dt_1\ldots dt_n$$

$$\int\limits_{-1}^{1}\cdots\int\limits_{-1}^{1}\prod_{j=2}^{n}L_j(t_j)\Big[\int\limits_{-1}^{1}L_1(t_1)\,\varphi\Big(x_1,x_2+\frac{x_nt_2}{2},\ldots,x_n+\frac{x_nt_n}{2}\Big)dt_1\Big]\times$$

$$\times dt_2 \ldots dt_n\Big|^p dx_1 \ldots dx_n\Big\}^{\frac{1}{p}}$$

$$= \Big\{ \int_{-\infty}^{\infty} \cdots \int_{-\infty}^{\infty} \int_{-1}^{1} \frac{1}{|x_n|^{p\alpha-\varepsilon}} \Big| \int_{-1}^{1} \cdots \int_{-1}^{1} \prod_{j=1}^{n} L_j(t_j)$$

$$\times \Big[\varphi\Big(x_1 + \frac{x_n t_1}{2}, \ldots, x_n + \frac{x_n t_n}{2}\Big)$$

$$- \varphi\Big(x_1, x_2 + \frac{x_n t_2}{2}, \ldots, x_n + \frac{x_n t_n}{2}\Big)\Big] dt_1 \ldots dt_n\Big|^p$$

$$\times dx_1 \ldots dx_n\Big\}^{\frac{1}{p}} \leqslant \int_{-1}^{1} \cdots \int_{-1}^{1} \prod_{j=1}^{n} |L_j(t_j)| \Big\{ \int_{-1}^{1} \frac{dx_n}{|x_n|^{p\alpha-\varepsilon}}$$

$$\times \int_{-\infty}^{\infty} \cdots \int_{-\infty}^{\infty} \Big| \varphi\Big(x_1 + \frac{x_n t_1}{2}, x_2 + \frac{x_n t_2}{2}, \ldots, x_n + \frac{x_n t_n}{2}\Big)$$

$$- \varphi\Big(x_1, x_2 + \frac{x_n t_2}{2}, \ldots, x_n + \frac{x_n t_n}{2}\Big)\Big|^p dx_1 \ldots dx_{n-1}\Big\}^{\frac{1}{p}} dt_1 \ldots dt_n \text{ [9]}$$

$$\leqslant \int_{-1}^{1} \cdots \int_{-1}^{1} \prod_{j=1}^{n} |L_j(t_j)| \Big\{ \int_{-1}^{1} \frac{dx_n}{|x_n|^{p\alpha-\varepsilon}}$$

$$\times \int_{-\infty}^{\infty} \cdots \int_{-\infty}^{\infty} \Big| \varphi\Big(u_1 + \frac{x_n t_1}{2}, u_2, \ldots, u_{n-1}, x_n + \frac{x_n t_n}{2}\Big)$$

$$- \varphi\Big(u_1, u_2, \ldots u_{n-1}, x_n + \frac{x_n t_n}{2}\Big)\Big|^p du_1 \ldots du_{n-1}\Big\}^{\frac{1}{p}} dt_1 \ldots dt_n.$$

In the right side of the last inequality we have put $u_1 = x_1$ and $u_i = x_i + x_n t_i/2, i = 2, \ldots, n-1$. By virtue of an imbedding theorm of Nikol'skiĭ ([3], Theorem 12, page 26), when $x_n = x_n^{(0)}$ is fixed the function φ belongs to the class $H_{p(n-1)}^{(\alpha-1/p)}(M_1)$, where

$$M_1 \leqslant c_1' M' + c_2' \|\varphi\|_p^{(n)}, \tag{5.7}$$

the nonnegative constants c_1' and c_2' not depending on f, M or $x_n^{(0)}$.

Since $0 < \alpha - 1/p < 1$ it follows from what has been said that

(9) We have applied the generalized Minkowski inequality here.

$$\left\{\int_{-\infty}^{\infty}\cdots\int_{-\infty}^{\infty}\left|\varphi\left(u_1+\frac{x_nt_1}{2},u_2,\ldots,u_{n-1},x_n+\frac{x_nt_n}{2}\right)-\varphi(u_1,u_2,\ldots,u_{n-1},x_n\right.\right.$$
$$\left.\left.+\frac{x_n t_n}{2}\right)\right|^p du_1\ldots du_{n-1}\Bigg\}^{\frac{1}{p}}\leqslant M_1\left|\frac{x_n t_1}{2}\right|^{\alpha-\frac{1}{p}}\leqslant M_1|x_n|^{\alpha-\frac{1}{p}}. \tag{5.8}$$

Substituting this expression into the estimate of I_3, we get

$$I_3\leqslant\int_{-1}^{1}\cdots\int_{-1}^{1}\prod_{j=1}^{n}|L_j(t_j)|\,dt_1\ldots dt_n\left\{\int_{-1}^{1}\frac{M_1^p|x_n|^{p\alpha-1}}{|x_n|^{p\alpha-\varepsilon}}dx_n\right\}^{\frac{1}{p}}$$
$$\leqslant L_sM_1\left\{\int_{-1}^{1}\frac{dx_n}{|x_n|^{1-\varepsilon}}\right\}^{\frac{1}{p}}=\frac{2L_sM_1}{\varepsilon^{\frac{1}{p}}},$$

where

$$L_s=\int_{-1}^{1}\cdots\int_{-1}^{1}\prod_{j=1}^{n}|L_j(t_j)|\,dt_1\ldots dt_n. \tag{5.9}$$

Finally, using (5.7) and (5.5), we conclude

$$I_3\leqslant\frac{b_1^{(s)}M+b_2^{(s)}\|f\|_p^{(n)}}{\varepsilon^{\frac{1}{p}}}. \tag{5.10}$$

In the second case we assume that the only kernel with the singular property (*) is L_n. Analogously to what was done above, we get

$$I_3\leqslant\left\{\int_{-\infty}^{\infty}\cdots\int_{-\infty}^{\infty}\int_{-1}^{1}\frac{1}{|x_n|^{p\alpha-\varepsilon}}\left|\int_{-1}^{1}\cdots\int_{-1}^{1}\prod_{j=1}^{n}L_j(t_j)\left[\varphi\left(x_1+\frac{x_n t_1}{2},\ldots,x_{n-1}+\frac{x_n t_{n-1}}{2},\right.\right.\right.\right.$$
$$\left.\left.x_n+\frac{x_n t_n}{2}\right)-\varphi\left(x_1+\frac{x_n t_1}{2},\ldots,x_{n-1}+\frac{x_n t_{n-1}}{2},x_n\right)\right]$$
$$\times\, dt_1\ldots dt_n\Big|^p dx_1\ldots dx_n\Bigg\}^{\frac{1}{p}}\leqslant\int_{-1}^{1}\cdots\int_{-1}^{1}\prod_{j=1}^{n}|L_j(t_j)|\left\{\int_{-1}^{1}\frac{dx_n}{|x_n|^{p\alpha-\varepsilon}}\right.$$
$$\times\int_{-\infty}^{\infty}\cdots\int_{-\infty}^{\infty}\left|\varphi\left(x_1+\frac{x_n t_1}{2},\ldots,x_{n-1}+\frac{x_n t_{n-1}}{2},x_n+\frac{x_n t_n}{2}\right)\right.$$
$$\left.\left.-\varphi\left(x_1+\frac{x_n t_1}{2},\ldots,x_{n-1}+\frac{x_n t_{n-1}}{2},x_n\right)\right|^p dx_1\ldots dx_{n-1}\right\}^{\frac{1}{p}}dt_1\ldots dt_n$$
$$=\int_{-1}^{1}\cdots\int_{-1}^{1}\prod_{j=1}^{n}|L_j(t_j)|\left\{\int_{-1}^{1}\frac{dx_n}{|x_n|^{p\alpha-\varepsilon}}\int_{-\infty}^{\infty}\cdots\int_{-\infty}^{\infty}\left|\varphi\left(u_1,\ldots,u_{n-1},x_n+\frac{x_n t_n}{2}\right)\right.\right.$$
$$\left.\left.-\varphi(u_1,\ldots,u_{n-1},x_n)^p\right|du_1\ldots du_{n-1}\right\}^{\frac{1}{p}}dt_1\ldots dt_n,$$

In the right side of the equality here we have put

$$u_i = x_i + \frac{x_n t_i}{2}, \quad i = 1, 2, \dots n-1.$$

We next note that by virtue of our assumptions $1/p < \alpha \leqslant 1$, which implies that in the case being considered by us the conditions of Theorem [4.1] are satisfied for $m = n-1$. Therefore, applying (4.3) with ϵ replaced by $\epsilon/2p$, we get

$$\int_{-\infty}^{\infty} \dots \int_{-\infty}^{\infty} \left| \varphi\left(u_1, \dots u_{n-1}, x_n + \frac{x_n t_n}{2}\right) - \varphi(u_1, \dots u_{n-1}, x_n) \right|^p du_1 \dots du_{n-1}$$

$$\leqslant \frac{\left(c_1'' M' + c_2'' \|\varphi\|_p^{(n)}\right)^p}{\varepsilon^p} \left| \frac{x_n t_n}{2} \right|^{p\alpha - 1 - \frac{\varepsilon}{2}} \leqslant \frac{\left(b_1' M + b_2' \|f\|_p^{(n)}\right)^p}{\varepsilon^p} |x_n|^{p\alpha - 1 - \frac{\varepsilon}{2}}.$$

We have used (5.5) here.

Substituting this expression into the estimate of I_3 and using the notation (5.9), we conclude

$$I_3 \leqslant \frac{L_s \left(b_1' M + b_2' \|f\|_p^{(n)}\right)}{\varepsilon} \left\{ \int_{-1}^{1} \frac{dx_n}{|x_n|^{1 - \frac{\varepsilon}{2}}} \right\}^{\frac{1}{p}} = \frac{b_1''^{(s)} M + b_2''^{(s)} \|f\|_p^{(n)}}{\varepsilon^{1 + \frac{1}{p}}}. \tag{5.11}$$

It follows from estimates (5.6), (5.10) and (5.11) that in the case $\alpha > 1/p$

$$I < \frac{\tilde{a}_1^{(s)} M + \tilde{a}_2^{(s)} \|f\|_p^{(n)}}{\varepsilon^{1 + \frac{1}{p}}}$$

for sufficiently small $\epsilon > 0$.

Now suppose $\alpha \leqslant 1/p$. In this case one should make use of a representation of the derivative $\partial^{s+\bar{r}} Kf / \partial x_1^{s_1} \cdots \partial x_n^{s_n}$ in the form of a linear combination of terms of form (3.17), viz.

$$\frac{1}{x_n^{s+1}} \int_{-1}^{1} \dots \int_{-1}^{1} \prod_{j=1}^{n} L_j(t_j) \varphi\left(x_1 + \frac{x_n t_1}{2}, \dots, x_n + \frac{x_n t_n}{2}\right) dt_1 \dots dt_n,$$

$$\varphi \in H_p^{(\alpha+1)}(M'), \ \|\varphi\|_p^{(n)} \leqslant M', \ M' \leqslant c_1 M + c_2 \|f\|_p^{(n)}.$$

We note that in the proof given above it was essential that $\varphi \in H_{p(n)}^{(r)}(M')$, where $r > 1/p$, which permitted us, by applying imbedding theorems, to consider the function φ on $(n-1)$-dimensional hyperplanes as a function of the class $H^{(r-1/p)}_{p(n-1)}$. On the other hand, the condition $r - 1/p \leqslant 1$ made it possible to apply Theorem [4.1]. Both circumstances obviously hold when $r = 1 + \alpha$ and $0 < \alpha \leqslant 1/p$. Therefore all of the arguments given above, as is easily seen, remain in force when α is replaced by $\alpha + 1$ and give the same estimate (5.11). The only change will be in the proof of inequality (5.8) when $\alpha = 1/p$. For in this case $\varphi \in H^{(1)}_{p(n-1)}(M_1)$. In order to estimate the norm of a first difference it suffices to note that $\varphi \in H^{(1-\epsilon)}_{p(n-1)}(cM_1/\epsilon)$. Hence

$$\left\{\int_{-\infty}^{\infty}\cdots\int_{-\infty}^{\infty}\left|\varphi\left(u_1+\frac{x_n t_1}{2}, u_2, \ldots, u_{n-1}, x_n+\frac{x_n t_n}{2}\right)\right.\right.$$

$$\left.\left.-\varphi\left(u_1, u_2, \ldots, u_{n-1}, x_n+\frac{x_n t_n}{2}\right)\right|^p du_1 \ldots du_{n-1}\right\}^{\frac{1}{p}} \leqslant \frac{c}{\varepsilon} M_1\left|\frac{x_n t_1}{2}\right|^{1-\varepsilon}$$

$$\leqslant \frac{c}{\varepsilon} M_1 |x_n|^{1+\alpha-\frac{1}{p}-\varepsilon},$$

and the use of this estimate will change the constants $b_1^{(s)}$ and $b_2^{(s)}$ of inequality (5.10). The theorem is proved.

REMARK. For derivatives not involving differentiation with respect to x_n we have the estimate

$$\left\|\frac{\partial^{s+\bar{r}} Kf}{\partial x_1^{s_1} \ldots \partial x_{n-1}^{s_{n-1}}}\right\|^{(n)}_{p[\,|x_n|,\, p(s-\alpha)+\varepsilon]} \leqslant \frac{a_1^{(s)} M + a_2^{(s)} \|f\|_p^{(n)}}{\varepsilon^{\frac{1}{p}}}, \tag{5.12}$$

in which the constants $a_1^{(s)}$ and $a_2^{(s)}$ do not depend on f, M or ϵ.

This follows from the fact that (see [3.5])

$$\frac{\partial^{s+\bar{r}} Kf(x)}{\partial x_1^{s_1} \ldots \partial x_{n-1}^{s_{n-1}}} = \frac{(-1)^s}{c x_n^s} \int_{-1}^{1} \cdots \int_{-1}^{1} K(t_n) \prod_{j=1}^{n-1} \frac{d^{s_j-\rho_j} K(t_j)}{dt_j^{s_j-\rho_j}}$$

$$\times f_{\xi_1^{\rho_1} \ldots \xi_{n-1}^{\rho_{n-1}}}\left(x_1+\frac{x_n t_1}{2}, \ldots, x_n+\frac{x_n t_n}{2}\right) dt_1 \ldots dt_n,$$

where $\rho_1 + \cdots + \rho_{n-1} = \bar{r}$ and $s_j \geqslant \rho_j$, $j = 1, 2, \ldots, n-1$, and hence only the kernels corresponding to the variables $t_1, \ldots, t_{n-1}$ can be a kernel with the singular property (*).

Inequality (5.12) follows from (5.6) and (5.10).

The following theorem corresponds to Theorem [5.1] in the case $p=\infty$.

[5.2]. *Suppose* $f \in H^{(r)}_{\infty(n)}(M)$, *with* $r=\bar{r}+\alpha$. *Then for any* $\epsilon>0$ *and* $s>0$

$$\left\| |x_n|^{s-\alpha+\varepsilon} \frac{\partial^{s+\bar{r}} Kf}{\partial x_1^{s_1}\dots\partial x_n^{s_n}} \right\|_\infty^{(n)} \leqslant a_1^{(s)} M + a_2^{(s)} \|f\|_p^{(n)}, \tag{5.13}$$

where the constants $a_1^{(s)}$ *and* $a_2^{(s)}$ *do not depend on* f *or* M. *When* $\alpha<1$ *the inequality continues to hold for* $\epsilon=0$.

PROOF. As in the proof of inequality (5.3), we will start from Theorem [3.6]. Suppose $\alpha<1$. It again suffices to estimate expressions of form (5.4) for $\varphi \in H^{(\alpha)}_{\infty(n)}(M')$. If, for example, L_i, $i=1,\dots,n$, is a kernel with the singular property (*), we have in terms of the notation (5.9)

$$\left| x_n^{s-\alpha} \frac{\partial^{s+\bar{r}} Kf}{\partial x_1^{s_1}\dots\partial x_n^{s_n}} \right|$$

$$= \Big| \frac{1}{x_n^\alpha} \int_{-1}^{1}\dots\int_{-1}^{1} \prod_{j=1}^{n} L_j(t_j) \Big[\varphi\Big(x_1+\frac{x_n t_1}{2},\dots, x_i+\frac{x_n t_i}{2},\dots, x_n+\frac{x_n t_n}{2}\Big)$$

$$-\varphi\Big(x_1+\frac{x_n t_1}{2},\dots, x_{i-1}+\frac{x_n t_{i-1}}{2}, x_i, x_{i+1}+\frac{x_n t_{i+1}}{2},\dots, x_n+\frac{x_n t_n}{2}\Big)\Big]$$

$$\times dt_1\dots dt_n \Big| \leqslant \frac{1}{|x_n|^\alpha} \int_{-1}^{1}\dots\int_{-1}^{1} \prod_{j=1}^{n} |L_j(t_j)|$$

$$\times \Big| \varphi\Big(x_1+\frac{x_n t_1}{2},\dots, x_i+\frac{x_n t_i}{2},\dots 1, x_n+\frac{x_n t_n}{2}\Big)$$

$$-\varphi\Big(x_1+\frac{x_n t_1}{2},\dots, x_i,\dots 1, x_n+\frac{x_n t_n}{2}\Big)\Big| dt_1\dots dt_n \leqslant \frac{L_s M'}{2} \leqslant a_1^{(s)} M + a_2^{(s)} \|f\|_p^{(n)}.$$

Taking the supremum of the left side of this inequality, we obtain (5.13) for $\epsilon=0$. If $\alpha=1$ and $\epsilon>0$ it suffices to note that $\varphi \in H^{(1-\epsilon)}_\infty(M'')$ and to carry out the same arguments as above.

[5.3]. *Suppose* $\rho>0$, $\rho=\bar{\rho}+\alpha$, *and suppose given* $\bar{\rho}+1$ *functions*

$$\psi_\lambda(x_1,\dots,x_{n-1}) \in H^{(\rho-\lambda)}_{p(n-1)}(M_\lambda),\ \lambda=0,1,\dots,\rho.$$

Then there exists a function f *defined on the whole space* E^n *and such that*

1°. $Kf \in H^{(\rho+1/p)}_{p(n)}(M)$, *where*

$$M \leqslant c_1 \sum_{\lambda=0}^{\rho} M_\lambda + c_2 \sum_{\lambda=0}^{\rho} \| \psi_\lambda \|_p^{(n-1)},$$

the constants c_1 *and* c_2 *not depending on* M_λ *or* ψ_λ, $\lambda = 0, 1, \cdots, \bar{\rho}$;

2°. $\|Kf\|_p^{(n)} \leqslant M$;

3°. $\partial^\lambda Kf/\partial x_n^\lambda{}_{x_n=0} = \psi_\lambda$, $\lambda = 0, 1, \cdots, \bar{\rho}$.

PROOF. We put $\varphi_0(x_1, \cdots, x_{n-1}) = \psi_0(x_1, \cdots, x_{n-1})$. Suppose now $\varphi_0, \cdots, \varphi_{\lambda-1}$ have been determined so that

a) $\varphi_k \in H_{p(n-1)}^{(\rho-k)}(M_k')$,

b) $\|\varphi_k\|_p^{(n)} \leqslant M_k'$, $k = 0, 1, \cdots, \lambda - 1$, where

$$M_k' \leqslant c_1^{(k)} \sum_{i=1}^{k} M_i + c_2^{(k)} \sum_{i=1}^{k} \| \psi_i \|_p^{(n-1)}, \tag{5.14}$$

We determine the function φ_λ from the equation

$$\psi_\lambda(x_1, \ldots, x_{n-1}) = \sum_{\lambda_1+\cdots+\lambda_n=\lambda,\ \lambda_n<\lambda} c_{\lambda_1\ldots\lambda_n} \frac{\partial^{\lambda_1+\cdots\lambda_{n-1}} \varphi_\lambda}{\partial x_1^{\lambda_1} \ldots \partial x_{n-1}^{\lambda_{n-1}}} + c_{0\cdots 0\lambda}\, \varphi_\lambda, \tag{5.15}$$

where the coefficients are uniquely determined by equality (4.15). All of the derivatives in the right side of (5.15) exist by virtue of condition a), and, what is more (see [3], Theorem 14, page 36),

$$\frac{\partial^{\lambda_1+\cdots+\lambda_{n-1}} \varphi_{\lambda_n}}{\partial x_1^{\lambda_1} \ldots \partial x_{n-1}^{\lambda_{n-1}}} \in H_{p(n-1)}^{(r)}(M_{\lambda_n}''), \tag{5.16}$$

where $r = (\rho - \lambda_n) - \Sigma_{i=1}^{n-1} \lambda_i = \rho - \lambda$ and

$$M_{\lambda_n}'' \leqslant c_1'' M_{\lambda_n}' + c_2'' \| \varphi_{\lambda_n} \|_p^{(n-1)}. \tag{5.17}$$

By virtue of the condition $c_{0\ldots 0\lambda} \neq 0$ (see (4.16)) equation (5.15) is solvable for φ_λ, it following from (5.16) and (5.14) that φ_λ satisfies conditions a) and b) for $k = \lambda$.

Further, according to a function extension theorem of Nikol'skiĭ ([4], page 87) there exists a function $f \in H_{p(n)}^{\rho+1/p}(M^*)$, where

$$M^* \leqslant a_1 \sum_{\lambda=0}^{\bar{\rho}} M'_\lambda + a_2 \sum_{\lambda=0}^{\bar{\rho}} \| \varphi_\lambda \|_p^{(n-1)},$$

such that

$$\| f \|_p^{(n)} \leqslant M^* \tag{5.18}$$

and

$$\frac{\partial^\lambda f(x_1, \ldots, x_{n-1}, 0)}{\partial x_n^\lambda} = \varphi_\lambda(x_1, \ldots, x_{n-1}). \tag{5.19}$$

Let us show that this is the desired function. From (5.19) we have

$$\frac{\partial^\lambda f(x_1, \ldots, x_{n-1}, 0)}{\partial x_1^{\lambda_1} \ldots \partial x_n^{\lambda_n}} = \frac{\partial^{\lambda_1 + \ldots + \lambda_{n-1}} \varphi_{\lambda_n}}{\partial x_1^{\lambda_1} \ldots \partial x_{n-1}^{\lambda_{n-1}}}, \quad \lambda_1 + \ldots + \lambda_n = \lambda.$$

But this implies according to (4.13) and (5.15) that

$$\frac{\partial^\lambda K f(x_1, \ldots, x_{n-1}, 0)}{\partial x_1^{\lambda_1} \ldots \partial x_n^{\lambda_n}} = \sum_{\lambda_1 + \ldots + \lambda_n = \lambda} c_{\lambda_1 \ldots \lambda_n} \frac{\partial^\lambda f(x_1, \ldots, x_{n-1}, 0)}{\partial x_1^{\lambda_1} \ldots \partial x_n^{\lambda_n}}$$

$$= \sum_{\lambda_1 + \ldots + \lambda_n = \lambda} c_{\lambda_1 \ldots \lambda_n} \frac{\partial^{\lambda_1 + \ldots + \lambda_{n-1}} \varphi_\lambda}{\partial x_1^{\lambda_1} \ldots \partial x_{n-1}^{\lambda_{n-1}}} = \psi_\lambda(x_1, \ldots, x_{n-1}).$$

Conditions 1° and 2° of Theorem [5.3] follow from Theorems [1.1] and [2.2].

We now formulate the main theorem.

[5.4] (FUNCTION EXTENSION THEOREM). *Suppose* $1 \leqslant p < \infty$, $\rho > 0$, $\rho = \bar{\rho} + \beta$, *where* $\bar{\rho}$ *is a nonnegative integer and* $0 < \beta \leqslant 1$, *and suppose given a function* $\psi_\lambda(x_1, \ldots, x_{n-1}) \in H_{p(n-1)}^{(\rho-\lambda)}(M_\lambda)$ *for each* $\lambda = 0, 1, \ldots, \rho$. *Suppose, further,* $r = \rho + 1/p = \bar{r} + \alpha$, *where* $\bar{r}$ *is a nonnegative integer and* $0 < \alpha \leqslant 1$. *Then there exists a function* $F(x_1, \ldots, x_n)$ *defined on* E^n *and satisfying the following conditions.*

1°. *All of the derivatives of* F *exist and are continuous everywhere except possibly on the hyperplane* $x_n = 0$.

2°. $F \in H_{p(n)}^{(r)}(M)$, *where*

$$M \leqslant c_1 \sum_{\lambda=0}^{\bar{\rho}} M_\lambda + c_2 \sum_{\lambda=0}^{\bar{\rho}} \| \psi_\lambda \|_p^{(n-1)},$$

the constants c_1 *and* c_2 *not depending on* M_λ *or the functions* ψ_λ, $\lambda = 0, 1, \ldots, \bar{\rho}$.

3°. $\|F\|_p^{(n)} \leqslant M.$

4°. $\partial^\lambda F(x_1, \ldots, x_{n-1}, 0)/\partial x_n^\lambda = \psi_\lambda(x_1, \ldots, x_{n-1}), \lambda = 0, 1, \ldots, \rho$.

5°. *For sufficiently small* $\epsilon > 0$ *and* $s = 1, 2, \ldots$

$$\left\| \frac{\partial^{s+\bar{r}} F}{\partial x_1^{s_1} \ldots \partial x_n^{s_n}} \right\|_{p[\,|x_n|,\, p(s-\alpha)+\epsilon]}^{(n)} \leqslant \frac{M^{(s)}}{\epsilon^{1+\frac{1}{p}}},$$

where

$$M^{(s)} \leqslant c_1^{(s)} \sum_{\lambda=0}^{\bar{\rho}} M_\lambda + c_2^{(s)} \sum_{\lambda=0}^{\bar{\rho}} \| \psi_\lambda \|_p^{(n-1)}.$$

In the case $p = \infty$ *it is only necessary to change the inequality in condition* 5° *to*

$$5^\infty. \qquad \left\| |x_n|^{s-\alpha+\epsilon} \frac{\partial^{s+\bar{r}} F}{\partial x_1^{s_1} \ldots \partial x_n^{s_n}} \right\|_\infty^{(n)} < \infty \quad (\alpha = \beta, \ \bar{r} = \bar{\rho}).$$

We remark that the existence of a function F satisfying conditions 2°–4° was proved by S. M. Nikol'skiĭ ([4], page 87). Thus the theorem is new only in regard to conditions 1°, 5° and 5^∞. Its proof follows directly from [5.1]–[5.3]; it suffices to take $F = Kf$, where f is the function constructed in [5.3].

We note that by way of addition to the above theorem at the present time Ja. S. Bugrov [17] has proved the possibility of an analogous extension from an $(n-1)$-dimensional hyperplane (and even an m-dimensional hyperplane for $1 \leqslant m \leqslant n-1$) onto the whole space in the case $\rho = 0$ if it is assumed that $H_p^{(0)} = L_p$. And in the case $p = 2$ and $n = 2$ he showed in the same paper that in extending a function from the unit circle under assumptions corresponding to our condition 5° it is possible to put $\epsilon = 0$. The question of this possibility in the general case remains open.

§6. Function extensions from a domain

Let G be a domain in E^n. For a function f given on G we introduce the notation

$$\| f \|_{p}^{(n)} = \{ \int_G |f|^p dG \}^{\frac{1}{p}}, \tag{6.1}$$

and, in the case of a weight function σ (see (5.1)),

$$\|\underset{G}{f}\|_{p(\sigma,\nu)}^{(n)} = \{\int \sigma^{\nu} |f|^p dG\}^{\frac{1}{p}}. \tag{6.2}$$

The set of all functions f such that $\|\underset{G}{f}\|_p^{(n)} < \infty$ is denoted by $L_p^{(n)}(G)$.

As before, when $G = E^n$ the index G will be dropped. We recall that the H-classes were defined by Nikol'skiĭ not only for functions given on the whole space E^n but also for functions given on bounded domains [4]. Let us present this definition. It will not be assumed that the domain G is finite.

For every domain $G \subset E^n$ and $\eta > 0$ we put $G_\eta = \{x\colon x \in G, \rho(x, E^n \backslash G) > \eta\}$, it being assumed in the case $G = E^n$ that $G_\eta = E^n$ for any $\eta > 0$.

A function f is said to be a function of class $H_{px_i}^{(r)}(M, G)$, $r = \bar{r} + \alpha$, if it together with its weak derivatives $\partial^k f/\partial x_i^k$, $k = 1, \cdots, \bar{r}$, is defined on G and satisfies the conditions

$$\left\|\underset{G}{\frac{\partial^k f}{\partial x_i^k}}\right\|_p^{(n)} < \infty, \quad k = 0, 1, \ldots r,$$

$$\left\|\underset{G_\eta}{\Delta_{x_i}^{(1)}\left(\frac{\partial^{\bar{r}} f}{\partial x_i^{\bar{r}}}, h\right)}\right\|_p^{(n)} < M |h|^\alpha \text{ when } \alpha < 1,$$

$$\left\|\underset{G_\eta}{\Delta_{x_i}^{(2)}\left(\frac{\partial^{\bar{r}} f}{\partial x_i^{\bar{r}}}, h\right)}\right\|_p^{(n)} \leqslant M |h| \text{ when } \alpha = 1,$$

for any h such that $0 < |h| < \eta$, where η is an arbitrary positive number. If a function f belongs to all of the classes $H_{px_i}^{(r_i)}(M_i, G)$, $i = 1, \cdots, n$, it is said to be a function of class $H_{p(n)}^{(r_1, \cdots, r_n)}(M_1, \cdots, M_n; G)$.

In particular, when $r_1 = \cdots = r_n = r$ and $M_1 = \cdots = M_n = M$ it is simply said to be a function of class $H_{p(n)}^{(r)}(M, G)$.

In this section we prove some auxiliary assertions.

[6.1]. *Suppose* $f \in H_{p(n)}^{(r_1, \cdots, r_n)}(G; M_1, \cdots, M_n)$. *Then for any* $\eta > 0$ *there exists a function* F *defined on* E^n *and satisfying the following conditions.*

1°. $F = f$ *on* G_η.

2°. $F \in H_{p(n)}^{(r_1, \cdots, r_n)}(M_1', \cdots, M_n')$, *where* $M_i' \leqslant c_{1\eta} M_i + c_{2\eta} \|f\|_p^{(n)}$.

3°. $\|F\|_p^{(n)} \leqslant c_\eta \|\underset{G}{f}\|_p^{(n)}$.

4°. *If* f *is a* k *times (infinitely) continuously differentiable function on* $G_\eta \backslash A$, *where* A *is a set such that* $\rho(A, E^n \backslash G_\eta) > 0$, *then* F *is* k *times (infinitely) continuously differentiable on* $E^n \backslash A$.

5°. $F = 0$ *outside* G.

In the case when G is a finite domain the existence of a function F satisfying conditions 1°–3° and 5° has been proved by Nikol'skii (see [4], Theorem 2.2, page 52). Let us indicate the method of proof in the general case. Suppose $\eta > 0$ is fixed and suppose under the presence of a set A (see condition 4°) $\eta' = \rho(A, E^n \backslash G_\eta)$. Put $\eta_0 = \min(\eta, \eta')$. We divide the whole space into cubes of diameter $\eta_0/3$ by means of, for example, the hyperplanes $x_i = m\eta_0/3\sqrt{n}$, $m = 0, \pm 1, \cdots$, $i = 1, \cdots, n$, and we number in some order the vertices of those cubes which intersect the closure $\overline{G}_\eta$, denoting them by P_k, $k = 1, 2, \cdots$. Consider the system of balls Q_k and Q'_k with centers at the points P_k and radii $\eta_0/3$ and $2\eta_0/3$ respectively. By virtue of the uniformity of the cubing there exists for any $\epsilon > 0$ a natural number $N_\epsilon > 0$ such that the ϵ-neighborhood of any point $P \in E^n$ intersects no more than N_ϵ balls Q_k. Suppose now $\psi(t)$ is an infinitely continuously differentiable function on the real line satisfying the conditions $\psi(t) = 0$ for $t \leqslant 1$ and $\psi(t) = 1$ for $t \geqslant 2$. Further, put $\psi_k(P) = \psi(3PP_k/\eta)$, where PP_k is the distance from an arbitrary point $P \in E^n$ to the point P_k, $k = 1, 2, \cdots$. Let

$$\Psi(x_1, \ldots, x_n) = \Psi(P) = 1 - \prod_{k=1}^{\infty} \psi_k(P). \tag{6.3}$$

For each point $P \in E^n$ there exist no more than a finite number of balls Q'_k containing it. Therefore the product in the right side of (6.3) contains only a finite number of terms different from unity (for example, no more than N_1) and hence the function Ψ is defined everywhere on E^n. Moreover, from the existence of the above mentioned number N_ϵ it follows that the function Ψ is infinitely continuously differentiable on E^n and that there exist constants c_λ, $\lambda = 0, 1, \cdots$, such that

$$\left| \frac{\partial^\lambda \Psi}{\partial x_1^{\lambda_1} \ldots \partial x_n^{\lambda_n}} \right| < c_\lambda \quad \text{for all} \quad \lambda_1 + \ldots + \lambda_n = \lambda.$$

Assuming now $f \equiv 0$ outside G, we put

$$F(P) = \Psi(P) f(P). \tag{6.4}$$

The function F is the desired function. The idea of the present construction is based on a construction of Whitney and Hestenes (see, for example, [29], §26.0, as well as [30, 31]).

The above method of uniform locally finite coverings will be repeatedly applied in the sequel. The proof of the fact that the function (6.4) satisfies conditions 2° and 3° is literally the same as the proof of the corresponding theorem of Nikol'skii ([4], loc. cit.), since it is based on only those properties of the functions f and Ψ which also hold in our case. The satisfaction of conditions 1° and 4° is obvious by virtue of the construction itself.

In conclusion we note that in the special case when G is a layer and $p = 2$ a similar question on the extension of a function has been studied by Nikol'skii ([5], page 253) but by another method.

REMARK. Theorem [6.1] permits one, for example, to completely (to within constants) carry over imbedding theorems ([3], page 26) to the case when $f \in H_{p(n)}^{(r_1,\dots,r_n)}(M_1, \cdots, M_n; G)$, where G is the layer $a < x_n < b$, if one is required to determine the H-class of a function f on a hyperplane $x_n = x_n^{(0)}$, $a < x_n^{(0)} < b$.

Suppose now Γ is either a bounded domain in the hyperplane $E^{n-1} = \{(x_i): x_n = 0\}$ or the whole hyperplane and

$$\Gamma_{+a}^{(n)} = \{(x_i) : (x_1, \dots, x_{n-1}) \in \Gamma,\ 0 < x_n < a \leqslant \infty\},$$

$$\Gamma_{-a}^{(n)} = \{(x_i) : (x_1, \dots, x_{n-1}) \in \Gamma,\ -a < x_n < 0\},\quad G = \Gamma_{+a}^{(n)} \cup \Gamma \cup \Gamma_{-a}^{(n)}.$$

Clearly, G is an n-dimensional domain.

[6.2]. *Suppose a function f together with its weak derivatives $\partial^{k_i} f/\partial x_i^{k_i}$ $(k_i = 1, \cdots, r_i; i = 1, \cdots, n)$ is defined on $\Gamma_{+a}^{(n)}$ and*

$$f \in L_p^{(n)}(\Gamma_{+a}^{(n)}),\quad \frac{\partial^{k_i} f}{\partial x_i^{k_i}} \in L_p^{(n)}(\Gamma_{+a}^{(n)}),\ k_i = 1, 2, \dots, r_i,\ i = 1, 2, \dots n.$$

Then there exists a function $F \in L_p^{(n)}(G)$ coinciding with f on $\Gamma_{+a}^{(n)}$, having the weak derivatives $\partial^{k_i} F/\partial x_i^{k_i}$ $(k_i = 1, \cdots, r_i; i = 1, \cdots, n)$ on G and such that

$$\| F\|_{p}^{(n)}{}_{G} \leqslant c \| f \|_{p}^{(n)}{}_{\Gamma_{+a}^{(n)}},\quad \left\| \frac{\partial^{k_i} F}{\partial x_i^{k_i}} \right\|_{p\,G}^{(n)} \leqslant c \left\| \frac{\partial^{k_i} f}{\partial x_i^{k_i}} \right\|_{p\,\Gamma_{+a}^{(n)}}^{n},\quad k_i = 1, 2, \dots, r_i, i = 1, 2 \dots n. \tag{6.5}$$

In the proof ot this theorem we again make use of an idea of Whitney and Hestenes ([29], §260). Suppose the numbers $\lambda_1, \cdots, \lambda_{r_{n+1}}$ have been determined from the system of equations

$$(-1)^k \lambda_1 + \left(-\frac{1}{2}\right)^k \lambda_2 + \ldots + \left(-\frac{1}{r_n+1}\right)^k \lambda_{r_n+1} = 1, \; k = 0, 1, 2, \ldots, r_n. \tag{6.6}$$

For $(x_1, \cdots, x_n) \in G$ we put

$$F(x_1, \ldots, x_n) = \begin{cases} f(x_1, \ldots, x_n) & \text{if } x_n > 0, \\ \sum_{k=1}^{r_n+1} \lambda_k f\left(x_1, \ldots, x_{n-1}, -\frac{x_n}{k}\right) & \text{if } x_n < 0. \end{cases} \tag{6.7}$$

Then the function $F(x_1, \cdots, x_n)$ effects the desired extension. A detailed proof of this assertion can be found in a paper of Nikol′skiĭ [44] as well as in the author's dissertation [43].

A similar result for the case when $f \in W_p^{(l)}$ (in the terminology of Sobolev [2]) has been obtained by Babič [32]. The construction (6.7) will be required in the sequel.

§7. Averagings relative to m-dimensional hyperplanes

In the present section we discuss the generalization of the results of §5 to the case of an arbitrary hyperplane and we give a general method of averaging functions with a variable averaging "radius."

The main contents of the present paper do not depend on the results of this section.

Let $r = \sqrt{x_{m+1}^2 + \cdots + x_n^2}$, $0 < m < n$, and let $\omega(r)$ be a continuously differentiable function of r such that $\lim_{r \to 0} \omega(r) = 0$, $\omega(r) \neq 0$ for $r \neq 0$, the mapping

$$u_i = x_i + t_i \omega(r), \; i = 1, 2, \ldots n, \tag{7.1}$$

is for any t_i, $|t_i| \leqslant 1, \cdots, n$, a one-to-one mapping of the "layer"

$$E_m = \{(x_i) : |x_i| < 1, \; i = m+1, \ldots n\}$$

onto a domain $G \subset E^n$ and

$$\left| \frac{\partial(u_1, \ldots, u_n)}{\partial(x_1, \ldots x_n)} \right| > A > 0 \text{ on } E_m. \tag{7.2}$$

Further, let

$$c = \left[\int_{-1}^{1} K(t)\,dt \right]^{n}, Q_{\omega} = \{(\xi_i) : |\xi_i - x_i| \leqslant \omega(r), \quad i = 1,\ 2, \dots\ n\}.$$

Suppose now

$$f \in L_p^{(n)},\ K[\omega(r),\ x,\ \xi] = K\left[\frac{x - \xi}{\omega(r)} \right]$$

(see §1) and

$$K_\omega f(x) = \frac{1}{c\omega^n(r)} \int_{x_1 - \omega(r)}^{x_1 + \omega(r)} \cdots \int_{x_n - \omega(r)}^{x_n + \omega(r)} \prod_{i=1}^{n} K[\omega(r), x_i, \xi_i]\, f(\xi_1, \dots, \xi_n)\, d\xi_1 \dots d\xi_n.$$

$$(7.3)$$

Making the change of variable $\xi_i = x_i + t_i \omega(r)$, $i = 1, \cdots, n$, here, we obtain

$$K_\omega f(x) = \frac{1}{c} \int_{-1}^{1} \cdots \int_{-1}^{1} \prod_{i=1}^{n} K(t_i) f[x_1 + t_1\omega(r), \dots, x_n + t_n\omega(r)]\, dt_1 \dots dt_n.$$

$$(7.4)$$

Analogously to (1.12) we have

$$\left\| K_\omega f \right\|_{\substack{(n)\\p \\ E_m}} \leqslant A^{-\frac{1}{p}} \left\| f \right\|_p^{(n)}.$$

$$(7.5)$$

For, applying the generalized Minkowski inequality and the substitution (7.1), we get

$$\left\| K_\omega f \right\|_{\substack{(n)\\p\\E_m}} = \left[\int_{-\infty}^{\infty} \cdots \int_{-\infty}^{\infty} dx_1 \dots dx_m \left| \int_{-1}^{1} \cdots \int_{-1}^{1} \frac{1}{c} \int_{-1}^{1} \cdots \int_{-1}^{1} \prod_{i=1}^{n} K(t_i) f(x_1 + t_1\omega(r), \dots) \right. \right.$$

$$\times dt_1 \dots dt_n \Big|^p \cdot dx_{m+1} \dots dx_n \Big]^{\frac{1}{p}} \leqslant \frac{1}{c} \int_{-1}^{1} \cdots \int_{-1}^{1} \prod_{i=1}^{n} K(t_i)$$

$$\times \left[\int_{-\infty}^{\infty} \cdots \int_{-\infty}^{\infty} \int_{-1}^{1} \cdots \int_{-1}^{1} |f(x_1 + t_1\omega(r), \dots)|^p dx_1 \dots dx_n \right]^{\frac{1}{p}} dt_1 \dots dt_n$$

$$\leqslant \frac{1}{c} A^{-\frac{1}{p}} \int_{-1}^{1} \cdots \int_{-1}^{1} \prod_{i=1}^{n} K(t_i)\, dt_1 \dots dt_n \left[\int_{-\infty}^{\infty} \cdots \int_{-\infty}^{\infty} |f(u_1, \dots, u_n)|^p\, du_1 \dots du_n \right]^{\frac{1}{p}}$$

$$= A^{-\frac{1}{p}} \left\| f \right\|_p^{(n)}.$$

Thus $K_\omega f \in L_p^{(n)}(E_m)$ if $f \in L_p^{(n)}$.

Further, as in [4.3], it can be shown that the existence of

$f(x_1, \cdots, x_m, 0, \cdots . 0) \in L_p^{(m)}$ in the sense of a mean value implies the existence of

$$K_\omega f(x_1, \ldots, x_m, 0, \ldots, 0) = f(x_1, \ldots, x_m, 0, \ldots, 0). \tag{7.6}$$

Finally, to formula (3.6) there corresponds in our case, as can be verified by a direct calculation, the formula

$$\left[\delta_j^i + \beta_i \frac{\omega'(r)}{\omega(r)} \frac{x_i}{r}(\xi_j - x_j)\right] K_{\xi_i}^{(j)} = -K_{x_j}^{(j)}, \; i, j = 1, 2, \ldots, n, \tag{7.7}$$

where $\beta_i = \Sigma_{j=m+1}^n \delta_j^i$ and for brevity we have written $K^{(j)} = K[\omega(r), \xi_j, x_j]$.

It follows from (7.7) that

$$\frac{\partial K_\omega f}{\partial x_i} = -\frac{1}{c\omega^n(r)} \sum_{j=1}^n \int \frac{\partial}{\partial \xi_j} \left\{\left[\delta_j^i + \beta_i \frac{\omega'(r)}{\omega(r)} \frac{x_i}{r}(\xi_j - x_j)\right] K^{(j)}\right\} \prod_{k \neq j} K^{(k)} f dQ_\omega. \tag{7.8}$$

In fact,

$$\frac{\partial K_\omega f}{\partial x_i} = \frac{1}{c\omega^n(r)} \sum_{j=1}^n \int \frac{\partial K^{(j)}}{\partial x_i} \prod_{k \neq j} K^{(i)} f dQ_\omega - \frac{n\beta_i \omega'(r) x_i}{c\omega^{(n+1)}(r) r} \int \prod_{j=1}^n K^{(j)} f dQ_\omega$$

$$= -\frac{1}{c\omega^n(r)} \sum_{j=1}^n \int \left[-K_{x_i}^{(j)} + \beta_i \frac{\omega'(r)}{\omega(r)} \frac{x_i}{r} K^{(j)}\right] \prod_{k \neq j} K^{(k)} f dQ_\omega$$

$$= -\frac{1}{c\omega^n(r)} \sum_{j=1}^n \int \left\{\left[\delta_j^i + \beta_i \frac{\omega'(r)}{\omega(r)} \frac{x_i}{r}(\xi_j - x_j)\right] K_{\xi_j}^{(j)} + \beta_i \frac{\omega'(r)}{\omega(r)} \frac{x_i}{r} K^{(j)}\right\}$$

$$\times \prod_{k \neq j} K^{(k)} f dQ_\omega = -\frac{1}{c\omega^n(r)} \sum_{j=1}^n \int \frac{\partial}{\partial \xi_j} \left\{\left[\delta_j^i + \beta_i \frac{\omega'(r)}{\omega(r)} \frac{x_i}{r}(\xi_j - x_j)\right] K^{(j)}\right\} \prod_{k \neq j} K^{(k)} f dQ_\omega.$$

From (7.8) it follows that under differentiation of the function $K_\omega f$ the order of a derivative relative to r is raised by either $1/\omega(r)$ or $\omega'(r)/\omega(r)$, depending on the properties of the function $\omega(r)$. In fact, making the change of variable $\xi_i = x_i + t_i\omega(r)$ in (7.8), we get

$$\frac{\partial K_\omega f}{\partial x_i} = \sum_{j=1}^n \frac{\delta_j^i}{c} \frac{1}{\omega(r)} \int_{-1}^1 \cdots \int_{-1}^1 \frac{\partial K(t_j)}{\partial t_j} \prod_{k \neq j} K(t_k) f dt_1 \ldots dt_n$$

$$+ \beta_i \frac{x_i}{r} \frac{\omega'(r)}{\omega(r)} \int_{-1}^1 \cdots \int_{-1}^1 \frac{d}{dt_j}[t_j K(t_j)] \prod_{k \neq i} K(t_k) f dt_1 \ldots dt_n.$$

Thus, if $\omega(r) = \eta r^k$, $k = 0, 1, \cdots$, the order of a derivative relative to r is raised by r^{-k}.

It can be shown that the averaging operator K_ω preserves the $H_p^{(r)}$-classes for $r < 1$. In order to obtain corresponding assertions for higher derivatives one should, of course, impose additional restrictions on the function $\omega(r)$.

The scheme indicated here of averaging functions with a variable averaging radius gives a quite general method of smoothing a function under preservation of its boundary values. It turns out in this connection to be possible to calculate the growth of the derivatives. The fact that the function $K_\omega f$ is generally determined only on the layer E_m is not an essential restriction since the results of the preceding section permit one to extend it from E_m onto all of E^n with preservation of its differential properties.

The consideration of the question in the large for an arbitrary, in a sense, function $\omega(r)$ leads to involved calculations. We therefore dwell only on a special case.

Let $\omega(r) = \eta r^2$, where the constant η has been chosen so that the mapping

$$u_i = x_i + \eta r^2 t_i \,(i = 1,2, \ldots, n), \quad 0 < \eta < 1, \tag{7.9}$$

has the following properties for any $|t_i| \leqslant 1$; it is a one-to-one mapping of the layer E_m into E^n and

$$\left|\frac{\partial (u_1, \ldots, u_n)}{\partial (x_1, \ldots, x_n)}\right| > \frac{1}{2}. \tag{7.10}$$

Such an η can always be chosen. For, when $\eta = 0$ the mapping (7.9) is the identity mapping, and therefore its Jacobian is equal to 1. But the Jacobian of this mapping continuously depends on η, and uniformly in $x_1, \cdots, x_n$ and $t_1, \cdots, t_n$. Hence there exists an $\eta_0 > 0$ such that condition (7.10) is satisfied for $0 < \eta < \eta_0$. Suppose now there exist a sequence $\eta_k \to 0$, $0 < \eta_k < \eta_0$, $k = 1, 2, \cdots$, and points $P^{(k)} = (x_1^{(k)}, \cdots, x_n^{(k)}) \in E_m$ and $Q^{(k)} = (y_1^{(k)}, \cdots, y_n^{(k)}) \in E_m$ that are mapped by the mapping

$$u_i = x_i + \eta_k r^2 t_i^{(k)} \tag{7.11}$$

into one and the same points. Then the points $P_1^{(k)} = (0, \cdots, 0, x_{m+1}^{(k)}, \cdots, x_n^{(k)})$ and $Q_1^{(k)} = (\xi_1^{(k)}, \cdots, \xi_m^{(k)}, y_{m+1}^{(k)}, \cdots, y_n^{(k)})$, where $\xi_i^{(k)} = y_i^{(k)} - x_i^{(k)}$, $i = 1, \cdots, m$, are also mapped by virtue of formulas (7.11) into one and the same points. Therefore

$$\xi_i^{(k)} + \eta_k (y_{m+1}^{2(k)} + \ldots + y_n^{2(k)}) t_i^{(k)} = \eta_k (x_{m+1}^{2(k)} + \ldots + x_n^{2(k)}) t_i^{(k)}$$

and hence $|\xi_i^{(k)}| \leqslant 2m\eta_0$, $i = 1, \cdots, m$, $k = 1, 2, \cdots$. Thus all of the points $P_1^{(k)}$ and $Q_1^{(k)}$ lie in the bounded parallelepiped $|x_i| \leqslant 2m\eta_0$, $i = 1, \cdots, m$, $|x_j| \leqslant 1$, $j = m + 1, \cdots, n$. When $\eta = 0$, as we have already noted, the mapping (7.9) converts into the identity mapping. Therefore, when $k \to \infty$, the distance between $P_1^{(k)}$ and $Q_1^{(k)}$ tends to zero. It can be assumed without loss of generality that the sequences of points $P_1^{(k)}$ and $Q_1^{(k)}$, $k = 1, 2, \cdots$, converge to a point P_0. By virtue of the implicit function theorem and the boundedness of the second derivatives of the mapping (7.9) the point P_0 has a fixed neighborhood on which (7.9) is a one-to-one mapping for all sufficiently small $\eta > 0$ and any t_i, $|t_i| \leqslant 1$, and which therefore cannot contain the points $P_1^{(k)}$ and $Q_1^{(k)}$.

The resulting contradiction proves our assertion.

We now put

$$K_2 f(x) = \frac{1}{c\eta^n r^{2n}} \int_{x_1-\eta r^2}^{x_1+\eta r^2} \cdots \int_{x_n-\eta r^2}^{x_n+\eta r^2} \prod_{j=1}^{n} K(\eta r^2, x_j, \xi_j) f(\xi_1, \ldots, \xi_n)\, d\xi_1 \ldots d\xi_n.$$

From (7.5) and (7.10) we then have

$$\|K_2 f\|_{p, E_m}^{(n)} \leqslant 2^{\frac{1}{p}} \|f\|_p^{(n)}. \tag{7.12}$$

Formula (7.7) converts into

$$\left(\delta_j^i + 2\beta_i x_i \frac{\xi_j - x_j}{r^2} \right) K_{\xi_j}^{(j)} = - K_{x_i}^{(j)};\ i, j = 1, 2, \ldots, n. \tag{7.13}$$

This implies the following more general formula (corresponding to (3.8)), which is obtained by a direct calculation:

$$\begin{aligned} &\frac{2\beta_{i_{s+1}} x_{i_{s+1}}}{r^2} \left(\frac{2(\xi_j - x_j)}{r^2} \right)^{\mu} \prod_{k=1}^{s} \left(\delta_j^{i_k} + 2\beta_{i_k} x_{i_k} \frac{\xi_j - x_j}{r^2} \right) K^{(j)} \\ &- \frac{\partial}{\partial x_{i_{s+1}}} \left[\left(\frac{2(\xi_j - x_j)}{r^2} \right)^{\mu} \prod_{k=1}^{s} \left(\delta_j^{i_k} + 2\beta_{i_k} x_{i_k} \frac{\xi_j - x_j}{r^2} \right) K^{(j)} \right] \\ &= \frac{\partial}{\partial \xi_j} \left[\left(\frac{2(\xi_j - x_j)}{r^2} \right)^{\mu} \prod_{k=1}^{s+1} \left(\delta_j^{i_k} + 2\beta_{i_k} x_{i_k} \frac{\xi_j - x_j}{r^2} \right) K^{(j)} \right] \\ &- \sum_{\nu=1}^{s} \beta_{i_\nu} \delta_{i_{s+1}}^{i_\nu} \left(\frac{2(\xi_j - x_j)}{r^2} \right)^{\mu+1} \prod_{\substack{k=1 \\ k \neq \nu}}^{s} \left(\delta_j^{i_k} + \beta_{i_k} x_{i_k} \right) \frac{\xi_j - x_j}{r^2} K^{(j)}, \\ &i_k,\ j = 1, 2, \ldots, n;\ s = 0, 1, \ldots;\ \mu = 0, 1, \ldots. \end{aligned} \tag{7.14}$$

A special characteristic of the multidimensional case $(n - m > 1)$ is the appearance here of the second term, which was not in (3.8).

As in §3, it can be shown that the function $K_2 f$ is infinitely differentiable on $E^n \backslash E^m$ when $f \in L_p^{(n)}$. In this connection we have the formula

$$\frac{\partial^s K_2 f(x)}{\partial x_{i_1} \cdots \partial x_{i_s}}$$

$$= \frac{1}{c\eta^n r^{2n}} \sum_{\lambda+\mu=s} a_{\lambda\mu\nu} \int \prod_{j=1}^{n} \frac{\partial^{\lambda_j}}{\partial \xi_j^{\lambda_j}} \left[\left(\frac{2(\xi_j - x_j)}{r^2} \right)^{\mu_j} \prod_{\nu_j}^{\lambda_j} \left(\delta_j^{i_{\nu_j}} + 2\beta_{i_{\nu_j}} \frac{\xi_j - x_i}{r^2} \right) K^{(j)} \right] f \, dQ_{r^2}, \tag{7.15}$$

Here $\lambda = \Sigma_1^n \lambda_j$, $\mu = \Sigma_1^n \mu_j$; $i_1, \cdots, i_s$ take values from 1 to n; each ν_j ranges over λ_j of the values $1, \cdots, s$, with $\nu_j \neq \nu_{j'}$ for $j \neq j'$, the notation $\Pi_{\nu_j}^{\lambda_j}$ meaning that the product is taken over the index ν_j and the number of factors is equal to λ_j; and $Q_{r^2} = \{(\xi_j): |\xi_j - x_j| \leqslant \eta r^2\}$. It can be shown that for fixed ν_j and $\lambda = s$

$$a_{\lambda\mu\nu} = (-1)^s;$$

more generally, the coefficients $a_{\lambda\mu\nu}$ depend on $\lambda_1, \cdots, \lambda_n, \mu_1, \cdots, \mu_n$, and the values of the ν_j $(j = 1, \cdots, n)$.

Finally, $\lambda > 0$ whenever $s > 0$.

To varying degrees, the other properties of the operator K studied by us earlier can be carried over to the operator K_2. As an application of the operator K_2, one can prove a theorem on function extensions from m-dimensional hyperplanes onto the whole space that is analogous to theorem [5.4]. We will not dwell on it here, since it will not be needed for our purposes and its presentation is quite cumbersome.

§8. Function extensions from smooth manifolds

A set $K^{(m)}$ lying in E^n will be called an *m-dimensional* $(1 \leqslant m \leqslant n - 1)$ *manifold* of order of smoothness k, more briefly *of smoothness* k, if it satisfies the following conditions.

1°. Every point $P_0 \in K^{(m)}$ has a neighborhood $U(P_0)$ in E^n for which there exists a one-to-one mapping of the set $U(P_0) \cap K^{(m)}$ onto an m-dimensional ball $Q^{(m)} = \{(u_i): u_1^2 + \cdots + u_m^2 < R^2\}$. The point of $Q^{(m)}$ corresponding to a point $P \in U(P_0) \cap K^{(m)}$ under this mapping will be denoted by $(u_1^{(P)}, \cdots, u_m^{(P)})$. The numbers $u_1^{(P)}, \cdots, u_m^{(P)}$ are called *local coordinates* of P (corresponding to the neighborhood $U(P_0)$ and the mapping of $U(P_0) \cap K^{(m)}$ onto $Q^{(m)}$), while the neighborhood $U(P_0)$ is called a *proper neighborhood* of P_0.

2°. Every point $P \in U^{(1)} \cap U^{(2)} \cap K^{(m)}$, where $U^{(1)}$ and $U^{(2)}$ are proper neighborhoods of points in $K^{(m)}$, has a neighborhood U such that $U \subseteq U^{(1)} \cap U^{(2)}$.

3°. For any two nondisjoint proper neighborhoods $U^{(1)}$ and $U^{(2)}$ a local coordinate system $(u_1^{(1)}, \cdots, u_m^{(1)})$ in the set $U^{(1)} \cap K^{(m)}$ can be expressed in terms of a local coordinate system $(u_1^{(2)}, \cdots, u_m^{(2)})$ in the set $U^{(2)} \cap K^{(m)}$ by means of k times continuously differentiable functions with a nonvanishing Jacobian.

4°. Every point $P_0 \in K^{(m)}$ has a proper neighborhood U such that if $(x_1^P, \cdots, x_n^P)$ denotes the coordinates in E^n of a point $P \in U \cap K^{(m)}$ and $(u_1^P, \cdots, u_m^P)$ denotes the local coordinates of P in a fixed local coordinate system, the functions

$$x_i^P = x_i^P(u_1^P, \ldots, u_m^P), \quad P \in U \cap K^{(m)}, \quad i = 1, 2, \ldots, n, \tag{8.1}$$

are k times continuously differentiable and the rank of the matrix $\|\partial x_i^P / \partial u_j^P\|$ $(i = 1, \cdots, n; j = 1, \cdots, m)$ is equal to m. The functions (8.1) are called a representation of $K^{(m)}$ in a neighborhood of P.

5°. Any two points of $K^{(m)}$ can be connected by a continuous path in $K^{(m)}$.

Property 4° means that in a certain neighborhood of each point $P \in K^{(m)}$ there exists a so-called explicit representation of $K^{(m)}$ relative to a certain set of $n - m$ coordinates expressed as functions of the remaining m coordinates. We thus require along with an intrinsic smoothness of the manifold (condition 3°) a certain smoothness of the imbedding of it in E^n. By a *submanifold* Γ of a manifold $K^{(m)}$ we will always mean a subset of $K^{(m)}$ that is connected and open relative to $K^{(m)}$.

Let $K^{(m)}$ be an m-dimensional manifold of smoothness $k \geqslant 1, P_0 \in K^{(m)}$; let $U = U(P_0)$ be a proper neighborhood of P_0 and $(u_1, \cdots, u_m)$ be a local coordinate system in $U \cap K^{(m)}$. The vectors $\overline{\tau}_i = \overline{\tau}_i(P) = (\partial x_1/\partial u_i, \cdots, \partial x_n/\partial u_i)$, $i = 1, \cdots, m, P \in U \cap K^{(m)}$, will be called tangent vectors to $K^{(m)}$ at P. Consider the system of equations

$$\alpha_1 \frac{\partial x_1}{\partial u_i} + \ldots + \alpha_n \frac{\partial x_n}{\partial u_i} = 0. \tag{8.2}$$

This system is linear and homogeneous, and at each point of $U \cap K^{(m)}$ the rank of the matrix $\|\partial x_i/\partial u_j\|$ is equal to m. Therefore every point $P \in U \cap K^{(m)}$ has a neighborhood $U(P)$ at each point of which there exists a system of $n - m$ orthogonal solutions $\overline{N}_j = (\alpha_1^{(j)}, \cdots, \alpha_n^{(j)}), j = m + 1, \cdots, n$, of (8.2) such that

1°. The functions $\alpha_i^{(j)} = \alpha_i^{(j)}(u_1, \cdots, u_m)$, $i = 1, \cdots, n$, $j = m+1, \cdots, n$, are $k-1$ times continuously differentiable in $U(P)$.

2°. $\Sigma_{i=1}^n (\alpha_i^{(j)})^2 = 1$, $j = m+1, \cdots, n$.

Clearly, $\bar{N}_{m+1}, \cdots, \bar{N}_n$ are unit vectors that are orthogonal to $K^{(m)}$.

Suppose now $k \geqslant 2$. Consider the mapping

$$x_i = x_i(u_1, \ldots, u_n) = x_i(u_1, \ldots, u_m) + \sum_{j=m+1}^{n} u_j \alpha_i^{(j)}, \quad i = 1, 2, \ldots, n. \tag{8.3}$$

It is $k-1$ times continuously differentiable and

$$J(u_1, \ldots, u_n) = \frac{\partial(x_1, \ldots, x_n)}{\partial(u_1, \ldots, u_n)}$$

$$= \begin{vmatrix} \dfrac{\partial x_1}{\partial u_1} + \sum\limits_{j=m+1}^{n} u_j \dfrac{\partial \alpha_1^{(j)}}{\partial u_1}, \ldots, \dfrac{\partial x_1}{\partial u_m} + \sum\limits_{j=m+1}^{n} u_j \dfrac{\partial \alpha_1^{(j)}}{\partial u_m}, & \alpha_1^{(m+1)}, \ldots, \alpha_1^{(n)} \\ \cdots\cdots\cdots\cdots\cdots\cdots\cdots\cdots & \cdots\cdots\cdots \\ \dfrac{\partial x_n}{\partial u_1} + \sum\limits_{j=m+1}^{n} u_j \dfrac{\partial \alpha_n^{(j)}}{\partial u_1}, \ldots, \dfrac{\partial x_n}{\partial u_m} + \sum\limits_{j=m+1}^{n} u_j \dfrac{\partial \alpha_n^{(j)}}{\partial u_m}, & \alpha_n^{(m+1)}, \ldots, \alpha_n^{(n)} \end{vmatrix},$$

which implies

$$J(u_1, \ldots, u_m, 0, \ldots, 0) = \begin{vmatrix} \dfrac{\partial x_1}{\partial u_1}, \ldots, \dfrac{\partial x_1}{\partial u_m}, \alpha_1^{(m+1)}, \ldots, \alpha_1^{(n)} \\ \cdots\cdots\cdots\cdots\cdots \\ \dfrac{\partial x_n}{\partial u_1}, \ldots, \dfrac{\partial x_n}{\partial u_m}, \alpha_n^{(m+1)}, \ldots, \alpha_n^{(n)} \end{vmatrix}.$$

The first m columns of this determinant are linearly independent since the rank of the matrix $\|\partial x_i/\partial u_j\|$ is by assumption equal to m, while the last $n-m$ columns are linearly independent and orthogonal to the first m columns by virtue of the way in which they were chosen. There therefore exists an $\eta > 0$ such that when $\Sigma_1^m (u_i - u_i^P)^2 + \Sigma_{m+1}^n u_j^2 < \eta$ we have

$$\left| \frac{\partial(x_1, \ldots, x_n)}{\partial(u_1, \ldots, u_n)} \right| > 0.$$

Applying the implicit function theorem, we finally obtain

[8.1]. *Every point P of an m-dimensional manifold $K^{(m)}$ of order of smoothness $k \geqslant 2$ has a neighborhood (relative to the manifold $K^{(m)}$) in which for any local coordinate system $(u_1, \cdots, u_m)$ there exists an $\epsilon > 0$ such that*

$$x_i = x_i(u_1, \ldots, u_m) + \sum_{j=m+1}^{n} u_j \alpha_i^{(j)}, \quad i = 1, 2, \ldots, n,$$

is a one-to-one $k-1$ *times continuously differentiable mapping with a Jacobian that does not vanish on the set*

$$W = \left\{(u_i): \sum_{i=1}^{m} (u_i - u_i^P)^2 < \epsilon^2,\ |u_j| < \epsilon, j = m+1, \cdots, n\right\}. \tag{8.4}$$

We now take an ϵ_1 such that $0 < \epsilon_1 < \epsilon$ and let

$$V_u = \{(u_i): (u_1, \ldots, u_m) \in V_u,\ |u_j| < \varepsilon,\ j = m+1, \ldots, n\}, \tag{8.5}$$

where V_u is an m-dimensional domain in the hyperplane $u_{m+1} = \cdots = u_n$ with boundary of measure zero such that

$$\mathring{V}_u \subseteq \left\{(u_i): \sum_{i=1}^{m} (u_i - u_i^P)^2 < \varepsilon_1^2,\ u_j = 0,\ j = m+1, \ldots, n\right\}. \tag{8.6}$$

Clearly, $\overline{V}_u \subset W$ (see (8.4)). By V and $\mathring{V}$ we respectively denote the images of V_u and $\mathring{V}_u$ under the mapping (8.3).

Every neighborhood V of the indicated type, with which there is associated a mapping of form (8.3) of a set V_u onto V, will be called a *canonical neighborhood* (of height ϵ_1), while the indicated mapping will be called a *canonical mapping.*

In the sequel we will use the notation

$$V_u^+ = \{(u_i): (u_1, \ldots, u_m) \in \mathring{V}_u,\ 0 < u_j < \varepsilon_1,\ j = m+1, \ldots, n\}, \tag{8.7}$$

and denote by V^+ the image of V_u^+ under the corresponding canonical mapping.

Under a canonical mapping a plane manifold $\mathring{V}_u$ is mapped onto a submanifold $\mathring{V}$ of the given manifold $K^{(m)}$.

A canonical mapping is a one-to-one $k-1$ times continuously and boundedly differentiable mapping, while the absolute value of its Jacobian is bounded from below by a positive constant:

$$\left|\frac{\partial (x_1, \ldots, x_n)}{\partial (u_1, \ldots, u_n)}\right| > A > 0.$$

If a canonical mapping takes a point $u = (u_1, \cdots, u_n)$ into a point $x = (x_1, \cdots, x_n)$, we can regard the numbers $u_1, \cdots, u_n$ as curvilinear coordinates of x in V, although we will also call them local coordinates, the coordinates of P

being denoted by $(u_1^P, \cdots, u_n^P)$.

Let P_0 be an arbitrary point of an m-dimensional manifold $K^{(m)}$. There then exists a neighborhood U of P_0 such that for every point $P \in U \cap K^{(m)}$ a system of $n-m$ unit vectors $\bar{N}_j = (\alpha_1^{(j)}, \cdots, \alpha_n^{(j)})$ that are orthogonal to each other and to the manifold $K^{(m)}$ can be chosen in such a way that the functions $\alpha_i^{(j)}(u_1, \cdots, u_m)$ $(i = 1, \cdots, n; j = m+1, \cdots, n)$ are $k-1$ times continuously differentiable functions of local coordinates. If this choice can be effected at each point P of $K^{(m)}$ in such a way that the normals $\bar{N}_j(P), j = m+1, \cdots, n$, are referenced in the same way, the manifold $K^{(m)}$ will be called an *orientable manifold* and the corresponding choice of normals will be said to be consistent.

In the sequel we will consider only orientable manifolds, without specifically saying so. We note only that every $(n-1)$-dimensional closed bounded manifold in E^n is orientable, and, if it is the boundary of a considered domain, a consistent choice of normals permits one to take inward or outward normals.

Let E be a subset of an m-dimensional manifold $K^{(m)}$ of smoothness $k \geqslant 2$ on which a consistent choice of normals has been made, and suppose E is covered by a finite number of canonical neighborhoods $V^{(\nu)}$ of height $\epsilon > 0$. This can always be done, for example, if $\bar{E}$ is compact and $\bar{E} \subseteq K^{(m)}$. In fact, if we choose a canonical neighborhood $V(P)$ of each point $P \in \bar{E}$, we can distinguish from the resulting covering of a compact set a finite subcovering $V^{(\nu)} = V(P^{(\nu)})$, where $V^{(\nu)}$ is the image under the corresponding canonical mapping of the set

$$V_{u(\epsilon_\nu)}^{(\nu)} = \left\{ (u_i^{(\nu)}) : \sum_{i=m+1}^{n} (u_i^{(\nu)} - u_i^{P(\nu)})^2 < \varepsilon_\nu^2, \ |u_j^{(\nu)}| < \varepsilon_\nu; \ j = m+1, \ldots, n \right\}$$

$$\nu = 1, 2, \ldots, \nu_0 .$$

Suppose $0 < \epsilon < \epsilon_\nu$. Then the system of canonical neighborhoods $V^{(\nu)}$, which are the images of the $V_{u(\epsilon)}^{(\nu)} = \{(u_i^{(\nu)}) \colon \Sigma_1^m (u_i^{(\nu)} - u_i^{P(\nu)})^2 < \epsilon^2, |u_j^{(\nu)}| < \epsilon; j = m+1, \cdots, n\}, \nu = 1, \cdots, \nu_0$, gives the desired covering of E. Let $\bar{\rho}(P)$ denote the radius vector of a point $P \in E^n$. The locus of the endpoints of the vectors

$$\bar{\rho} = \bar{\rho}(P) + \sum_{j=m+1}^{n} v_j \bar{N}_j(P), \ |v_j| < \eta < \varepsilon \ (j = m+1, \ldots, n), P \in E_j, \tag{8.8}$$

is called the cylindroid of height η constructed on E and is denoted by $K_\eta^{(n)}(E)$, while the loci of the endpoints of the vectors

$$\bar{\rho}(P) + \sum_{\substack{i=m+1 \\ i \neq j}}^{n} v_i \bar{N}_i(P) + \eta \bar{N}_j; \ |v_i| < \eta \tag{8.9}$$

$$(i = m+1, \ldots, j-1, \ j+1, \ldots, n), \quad P \in E,$$

$$\bar{\rho}(P) + \sum_{\substack{i=m+1 \\ i \neq j}}^{n} v_i \bar{N}_i(P) - \eta \bar{N}_j; \ |v_i| < \eta \tag{8.10}$$

$$(i = m+1, \ldots, j-1, \ j+1, \ldots, n), \quad P \in E,$$

$$\bar{\rho}(P) + \sum_{i=m+1}^{n} v_i \bar{N}_i(P); \ |v_i| < \eta \ (i = m+1, \ldots, j-1, \ j+1, \ldots, n), \quad v_j = 0, \ P \in E, \tag{8.11}$$

are respectively an upper base $K_{+\eta j}^{(n-1)}(E)$, a lower base $K_{-\eta j}^{(n-1)}(E)$ and a middle section $K_{0j}^{(n-1)}(E)$. A half-cylindroid $K_{+\eta j}^{(n)}(E)$ is the locus of the endpoints of the vectors

$$\bar{\rho}(P) + \sum_{i=m+1}^{n} v_i \bar{N}_i(P), \tag{8.12}$$

$$|v_i| < \eta \quad (i = m+1, \ldots, j-1, \ j+1, \ldots, n), \ 0 < v_j < \eta, \ P \in E$$

The half-cylindroids $K_{-\eta j}^{(n)}(E), j = m+1, \cdots, n$, are defined analogously. When $m = n-1$ and $j = n$ we will drop the index j.

If the set E is open in $K^{(m)}$, the set $K_\eta^{(n)}(E)$ is open in E^n and hence a neighborhood of E. If E itself is an m-dimensional manifold the sets $K_{+\eta j}^{(n-1)}(E)$, $K_{-\eta j}^{(n-1)}(E)$ and $K_{0j}^{(n-1)}(E)$ are $(n-1)$-dimensional manifolds of order $k-1$, being in the case $m = n-1$ homeomorphic to E and obtainable from each other by a continuous deformation with respect to the normals (in this case E coincides with $K_0^{(n-1)}(E)$).

For each endpoint $P \in K_\eta^{(n)}(E)$ of the vector $\bar{\rho}(P_0) + \Sigma_{j=m+1}^n v_j^P \bar{N}_j, P_0 \in E$, we put

$$r_v(P) = \sqrt{\sum_{j=m+1}^{n} (v_j^P)^2}$$

and call it the *distance* from P to E (or to $K^{(m)}$) with respect to the normals

$\overline{N}_{m+1}, \cdots, \overline{N}_n$. In going over to another consistent system of normals in the case of a fixed compact set E the function $r_v(P)$ varies by a function that is bounded in absolute value.

Suppose Γ is a submanifold of $K^{(m)}$, $\overline{\Gamma}$ is a compact set, $\overline{\Gamma} \subseteq K^{(m)}$, $P \in K_\eta^{(n)}(\Gamma)$ and $r(P) = \rho(P, \Gamma)$. Then for any consistent choice of normals there exist constants $c'_v > 0$ and $c''_v > 0$ such that

$$c'_v r(P) \leqslant r_v(P) \leqslant c''_v r(P). \tag{8.13}$$

In the sequel we will only be interested in the order of $r(P)$ for $P \to \Gamma$. We will therefore not distinguish between $r(P)$ and $r_v(P)$.

Suppose that on a manifold $K^{(m)}$ a point function f is given, V is a canonical neighborhood, $(u_1, \cdots, u_n)$ is a local coordinate system in V and V_u is the preimage of V under the corresponding canonical mapping. We will denote by f_u the function defined on $\overset{\circ}{V}$ (see (8.6)) by the equality

$$f_u(u_1^P, \ldots, u_m^P) = f(P), \quad P \in \overset{\circ}{V}. \tag{8.14}$$

Suppose now two functions f and g are given on $K^{(m)}$. We say that these functions are equivalent if for any canonical neighborhood V with a local coordinate system $(u_1, \cdots, u_n)$ we have

$$f_u(u_1^P, \ldots, u_m^P) = g_u(u_1^P, \ldots, u_m^P), \quad P \in V \cap K^{(m)},$$

almost everywhere on $\overset{\circ}{V}_u$.

A subset $E \subseteq K^{(m)}$ will be said to be of zero measure if for any canonical neighborhood V the set $E_u = \{(u_i^P): P \in E \cap V\}$ has an m-dimensional measure equal to zero.

In the sense of this definition equivalent functions f and g given on $K^{(m)}$ coincide on $K^{(m)}$ almost everywhere. We will not distinguish between equivalent functions in the sequel.

We will say that a function f defined on $K^{(m)}$ is of class $H_{p(m)}^{(r)}(M, K^{(m)})$ if there exists a system $\mathfrak{S} = \{V\}$ of canonical neighborhoods forming a locally finite covering of $K^{(m)}$ such that for any neighborhood $V \in \mathfrak{S}$ (see §6, (8.6) and (8.14))

$$f_u \in H_{p(m)}^{(r)}(M, \overset{\circ}{V}_u).$$

This definition is a natural one since by virtue of the corresponding results of Nikol'skiĭ ([4], page 59) the H-classes of functions remain invariant in a sense under sufficiently smooth transformations of the variables, which is guaranteed in our case by requirement 3° in the definition of a manifold. We will not dwell here on the question of the dependence of the H-class of a function on the choice of the system $\mathfrak{S}$, which will always be assumed to be fixed. We only note that in regard to a class $H^{(r)}_{p(m)}(M, K_1^{(m)})$, where $K_1^{(m)}$ is a submanifold of $K^{(m)}$ such that $\overline{K}_1^{(m)}$ is compact and $\overline{K}_1^{(m)} \subset K^{(m)}$ (only these cases will be encountered in the applications considered by us), the passage from $\mathfrak{S}$ to another system affects only the value of the constant M (assuming that the degree of smoothness of $K^{(m)}$ satisfies the condition $k \geqslant \bar{r} + 2$). This is true, for example, when $K_1^{(m)} = K^{(m)}$ is a closed bounded manifold; the stated assertion was proved for this case under assumptions not essentially different from our own by Nikol'skiĭ ([4], §4).

Suppose $\mathfrak{S} = \{V\}$ is a system of canonical neighborhoods forming a locally finite covering of a manifold $K^{(m)}$. We denote by $A^{(\nu)}, \nu = 1, 2, \cdots$, the members of a family of sets satisfying the following conditions:

1°. Each $A^{(\nu)}$ is entirely contained in every canonical neighborhood $V \in \mathfrak{S}$ with which it has a nonempty intersection.

2°. The intersection of two different $A^{(\nu)}$ is of measure zero.

3°. $K^{(m)} = \bigcup_\nu A^{(\nu)}$.

A choice of such $A^{(\nu)}$ can always be effected for any system $\mathfrak{S}$. If $A^{(\nu)} \subseteq V$, we denote by u_ν the canonical mapping corresponding to V, and by $A_u^{(\nu)}$ the pre-image of $A^{(\nu)}$ under u_ν. Suppose, finally, that a function f is given on $K^{(m)}$.

We put

$$\|f\|^{(m)}_{p\,K^{(m)}(\mathfrak{S})} = \sum_\nu \|fu_\nu\|^{(m)}_{p\,A_u^{(\nu)}}. \tag{8.15}$$

It is also possible to define in a natural way an invariant norm of f that does not depend on a choice of local representations of the manifold $K^{(m)}$, i.e. on the choice of a system $\mathfrak{S}$; namely, one can put

$$\|f\|^{(m)}_{p\,K^{(m)}} = \left\{\int |f|^p dK^{(m)}\right\}^{\frac{1}{p}}, \tag{8.16}$$

where $dK^{(m)}$ is a volume element of $K^{(m)}$.

In the case when $K_1^{(m)}$ is a submanifold of $K^{(m)}$ such that $\overline{K}_1^{(m)}$ is compact and $\overline{K}_1^{(m)} \subset K^{(m)}$, the passage from a system $\mathfrak{S}$ to another system causes a finite variation in the norm $\|f\|^{(m)}_{p\,K^{(m)}(\mathfrak{S})}$. In addition, for a system there exist constants c_1 and

c_2 (depending on $\mathfrak{S}$ but not on f) such that

$$c_1 \|f\|_{p}^{(m)} \leqslant \|f\|_{p}^{(m)} \leqslant c_2 \|f\|_{p}^{(m)},$$
$$\quad K_1^{(m)} \qquad K_1^{(m)}(\mathfrak{S}) \qquad K_1^{(m)}$$

Suppose further that G is a domain in E^n, $K^{(m)}$ is a manifold of order $k \geqslant 2$, $K^{(m)} \subseteq G$, a function f is given on G and V is a canonical neighborhood with a local coordinate system $(u_1, \cdots, u_n)$. We put

$$f_u(u_1^P, \ldots, u_n^P) = f(P), \quad P \in V. \tag{8.17}$$

We will say that a function f takes the value $\varphi = f|_{K^{(m)}}$ on a manifold $K^{(m)}$ (or, more precisely, tends to the value φ) in the sense of convergence in the mean (for a given p, $1 \leqslant p < \infty$) if for every point $P_0 \in K^{(m)}$ there exists a canonical neighborhood $V \ni P_0$ such that

$$\lim_{\sum_{m+1}^{n} u_i^2 \to 0} \int \cdots \int_{\overset{\circ}{V}_u} |f_u(u_1, \ldots, u_n) - \varphi_u(u_1, \ldots, u_m)|^p du_1 \cdots du_m = 0. \tag{8.18}$$

This definition is a natural one, since property (8.18) is preserved under appropriate restrictions on f for mappings of the independent variables with a bounded Jacobian that are invariant relative to the hyperplane $u_{m+1} = \cdots = u_n = 0$, which in the case under consideration is clearly true when going over from one local coordinate system to another. In particular, it can be shown that the function $f|_{K^{(m)}}$ defined in this way does not depend on a consistent choice of normals $\bar{N}_{m+1}, \cdots, \bar{N}_n$ on $K^{(m)}$. This was proved under assumptions not essentially different from our own by Nikol'skiĭ ([4], pages 92–98).

It follows from what has been said that $f|_{K^{(m)}}$ exists in the above defined sense if $f \in H_{p(n)}^{(r)}(M, G)$, where $r > (n-m)/p$ and $K^{(m)} \subset G$. This follows directly from an imbedding theorem of Nikol'skiĭ ([3], page 26).

Suppose now that for every canonical neighborhood V all of the weak derivatives $\partial^\lambda f_u / \partial u_{m+1}^{\lambda_{m+1}} \cdots \partial u_n^{\lambda_n}$ of order $\lambda = \lambda_{m+1} + \cdots + \lambda_n$ exist in $V \backslash K^{(m)}$. We put by definition

$$\frac{\partial^\lambda f(P)}{\partial N_{m+1}^{\lambda_{m+1}} \cdots \partial N_n^{\lambda_n}} = \frac{\partial^\lambda f_u(u_1^P, \ldots, u_n^P)}{\partial u_{m+1}^{\lambda_{m+1}} \cdots \partial u_n^{\lambda_n}} \tag{8.19}$$

and

$$\left.\frac{\partial^\lambda f(P)}{\partial N_{m+1}^{\lambda_{m+1}}\cdots\partial N_n^{\lambda_n}}\right|_{K^{(m)}\cap V}=\frac{\partial^\lambda f\left(u_1^P,\ldots,u_m^P,0,\ldots 0\right)}{\partial u_{m+1}^{\lambda_{m+1}}\cdots\partial u_n^{\lambda_n}} \tag{8.20}$$

(if, of course, the expression on the right exists in the sense of (8.18)).

In the case when the order of smoothness k of $K^{(m)}$ is such that $k\geqslant\lambda+1$ and $f\in H_{p(n)}^{(r)}(M)$, where $r-\lambda-(n-m)/p>0$, the normal derivatives defined in the above way transform under a passage from one local coordinate system to another (in particular, under another consistent choice of normals) according to the usual classical formulas. This was obtained in the same paper of Nikol'skiĭ ([4], pages 101–104) under assumptions not essentially (for carrying out the proof) different from our own.

Let G be a domain in $E^m\subset E^n$, $1\leqslant m<n$, $p>0$, $\rho=\bar\rho+\beta$, and for each $l=0,1,\cdots,\bar\rho$ suppose given all possible systems of nonnegative integers $\lambda_{m+1}^{(l)},\cdots,\lambda_n^{(l)}$ such that $\Sigma_{i=m+1}^n\lambda_i^{(l)}=l$. Further, suppose given in G for each system $\lambda_{m+1}^{(l)},\cdots,\lambda_n^{(l)}$ a function $\psi_{\lambda_{m+1}^{(l)},\cdots,\lambda_n^{(l)}}\in H_{p(m)}^{(\rho-l)}(M_{\lambda_{m+1}^{(l)}\cdots\lambda_n^{(l)}},G)$, where $r=\rho+(n-m)/p=\bar r+\alpha$, and suppose given a system of nonnegative functions

$$\varphi_s(r),\ s=1,2,\ldots,s_0\leqslant\infty,\ r=\sqrt{u_{m+1}^2+\cdots+u_n^2},$$

defined on $E_m=\{(u_i)\colon (u_i)\in E^n;|u_j|\leqslant 1, j=m+1,\cdots,n\}$. We say that the system of functions $\{\psi_{\lambda_{m+1}^{(l)}\cdots\lambda_n^{(l)}}\}$ is φ_s extendable from G if for any $\eta>0$ there exists a function F defined on E^n satisfying the following conditions.

1°. F is infinitely continuously differentiable everywhere on E^n except possibly the hyperplane

$$E^m=\{(u_i)\colon u_{m+1}=\ldots=u_n=0\}.$$

2°. $F\in H_{p(n)}^{(r)}(M)$, where

$$M\leqslant c_1\Sigma M_{\lambda_{m+1}^{(l)}\cdots\lambda_n^{(l)}}+c_2\Sigma\|\psi_{\lambda_{m+1}^{(l)}\cdots\lambda_n^{(l)}}\|_p^{(k)},$$

the constants c_1 and c_2 being independent of $\psi_{\lambda_{m+1}^{(l)}\cdots\lambda_n^{(l)}}$ and $M_{\lambda_{m+1}^{(l)}\cdots\lambda_n^{(l)}}$.

3°. $\|F\|_p^{(n)}\leqslant M$.

4°. $$\frac{\partial^{\lambda_{m+1}^{(l)}+\ldots+\lambda_n^{(l)}}F(u_1,\ldots,u_m,0,\ldots,0)}{\partial u^{\lambda_{m+1}^{(l)}}\cdots\partial u^{\lambda_n^{(l)}}}=\psi_{\lambda_{m+1}^{(l)}\cdots\lambda_n^{(l)}}$$

for $(u_1, \cdots, u_m) \in G_\eta = \{P: \rho(P, E^m \backslash G) > \eta\}$.

$$5°. \qquad \int_{E^n} \cdots \int \varphi_s(r) \left| \frac{\partial^{\bar{r}+s} F}{\partial u_1^{s_1} \cdots \partial u_n^{s_n}} \right|^p du_1 \cdots du_n < \infty,$$

where $s_1 + \cdots + s_n = \bar{r} + s$ and $s = 1, 2, \cdots, s_0$.

If the above assumptions hold with the exception that the functions $\psi_{\lambda_{m+1}^{(l)} \cdots \lambda_n^{(l)}}$ are given not on a flat manifold but on an arbitrary manifold $K^{(m)}$, with

$$\psi_{\lambda_{m+1}^{(l)} \cdots \lambda_n^{(l)}} \in H_{p(m)}^{(\rho - l)} (M_{\lambda_{m+1}^{(l)} \cdots \lambda_n^{(l)}}, K^{(m)})$$

and if for any canonical neighborhood $V \in \mathfrak{S}$ which is the image of V_u under a canonical mapping $x = x(u)$ the functions $\psi_{\lambda_{m+1}^{(l)} \cdots \lambda_n^{(l)}}(x(u))$ are φ_s extendable from $\overset{\circ}{V}_u$ (see (8.6)), we will say that the system $\{\psi_{\lambda_{m+1}^{(l)} \cdots \lambda_n^{(l)}}\}$ is φ_s extendable from $K^{(m)}$.

Thus the requirement of φ_s extendability in the case of an arbitrary manifold $K^{(m)}$ means in a sense a corresponding local extendability of the considered functions. The purpose of the present section is a proof of the fact that local φ_s extendability of a system $\{\psi_{\lambda_{m+1}^{(l)} \cdots \lambda_n^{(l)}}\}$ implies its extendability in the large for any φ_s, $s = 1, 2, \cdots$.

As an example, we note

[8.2]. *Let G be a domain in the $(n-1)$-dimensional hyperplane $E^{n-1} = \{(x_i): x_n = 0\}$. Then the system of functions $\psi_\lambda(x_1, \cdots, x_{n-1}) \in H_{p(n-1)}^{(\rho-\lambda)}(M_\lambda, G)$, $\rho = \bar{\rho} + \beta$, $\lambda = 0, 1, \cdots, \bar{\rho}$, is $|x_n|^{p(s-\alpha)+\epsilon}$ extendable from G, where $r = \rho + 1/p = \bar{r} + \alpha$, $\epsilon > 0$, $s = 1, 2, \cdots$.*

In fact, it follows from Theorem [6.1] that for any $\eta > 0$ the functions ψ_λ can be extended from the domain G_η onto all of E^{n-1} as functions $\tilde{\psi}_\lambda$ such that $\tilde{\psi}_\lambda = \psi_\lambda$ on G_η and $\tilde{\psi}_\lambda \in H_{p(n-1)}^{(\rho-\lambda)}(\tilde{M}_\lambda)$, where

$$\tilde{M}_\lambda \leqslant c_{1\eta} M_\lambda + c_{2\eta} \|\psi_\lambda\|_p^{(n-1)}, \quad \|\tilde{\psi}_\lambda\|_p^{(n-1)} \leqslant c_\eta \|\psi_\lambda\|_p^{(n-1)}.$$

Applying Theorem [5.4] to the system of functions $\{\tilde{\psi}_\lambda\}$, $\lambda = 0, 1, \cdots, \bar{\rho}$, we obtain the function F required in the definition of the φ_s extendability of a system of functions when $\varphi_s(r) = |x_n|^{p(s-\alpha)+\epsilon}$.

[8.3]. (THEOREM ON THE φ_s EXTENSION OF FUNCTIONS IN THE LARGE). *Suppose*

$1 \leqslant m < n$, $1 \leqslant p \leqslant \infty$, $K^{(m)}$ *is an m-dimensional manifold of order of smoothness* $k \geqslant 2$, O *is an arbitrary neighborhood of it*, $\rho > 0$, $\rho = \bar{\rho} + \beta$, $r = \rho + (n - m)/p = \bar{r} + \alpha$, $\bar{r} + 2 \leqslant k$ *and* $\varphi_s(t)$ *is a sequence of nonnegative measurable functions defined on the interval* $[0, t_0]$, $t_0 > 0$, $s = 1, \cdots, k - \bar{r} - 1$. *Further, for each* $l = 0, 1, \cdots, \bar{\rho}$ *suppose given all possible systems of nonnegative integers* $\lambda^{(l)}_{m+1}, \cdots, \lambda^{(l)}_n$ *such that* $\Sigma^n_{j=m+1}\lambda^{(l)}_j = l$, *and for each admissible system* $\lambda^{(l)}_{m+1}, \cdots, \lambda^{(l)}_n$ *suppose given a function*

$$\psi_{\lambda^{(l)}_{m+1}\cdots\lambda^{(l)}_n} \in H^{(\rho-l)}_{p(m)}(M_{\lambda^{(l)}_{m+1}\cdots\lambda^{(l)}_n}, K^{(m)}),$$

the system of functions $\{\psi_{\lambda^{(l)}_{m+1}\cdots\lambda^{(l)}_n}\}$ *being* φ_s *extendable. Finally, suppose* $G = E^n \setminus (\bar{K}^{(m)} \setminus K^{(m)})$ *(clearly, G is an open set),* $\mathfrak{G}$ *is a domain in* E^n *such that* $\bar{\mathfrak{G}}$ *is compact and* $\bar{\mathfrak{G}} \subset G$, $G_K = \mathfrak{G} \cap K^{(m)}$ *and* $r(P)$ *is the distance from a point* P *to* $K^{(m)}$ *"with respect to the normals" to* $K^{(m)}$. *Then there exists a function* $F = F(x_1, \cdots, x_n)$ *defined on* E^n *and satisfying the following requirements.*

1°. *F is continuously differentiable on* $E^n \setminus \bar{K}^{(m)}$ *up to order* $k - 1$ *inclusively.*

2°. $F \in H^{(r)}_{p(n)}(M, \mathfrak{G})$, *where*

$$M \leqslant c_1 \Sigma M_{\lambda^{(l)}_{m+1}\cdots\lambda^{(l)}_n} + c_2 \Sigma \|\psi_{\lambda^{(l)}_{m+1}\cdots\lambda^{(l)}_n}\|^{(m)}_p,$$

the constants c_1 *and* c_2 *not depending on* $M_{\lambda^{(l)}_{m+1}\cdots\lambda^{(l)}_n}$ *or* $\psi_{\lambda^{(l)}_{m+1}\cdots\lambda^{(l)}_n}$ *but generally depending on the choice of the domain* $\mathfrak{G}$.

3°. $\|F\|^{(n)}_{\mathfrak{G}\, p} \leqslant M$.

4°.
$$\left.\frac{\partial^l F}{\partial N^{\lambda^{(l)}_{m+1}}_{m+1} \cdots \partial N^{\lambda^{(l)}_n}_n}\right|_{K^{(m)}} = \psi_{\lambda^{(l)}_{m+1}\cdots\lambda^{(l)}_n}.$$

5°. *There exists an* $\eta = \eta(\mathfrak{G}) > 0$ *such that*

$$\int_{K^{(n)}_\eta(\mathfrak{G}_k)} \varphi_s[r(P)] \left|\frac{\partial^{\bar{r}+s} F(P)}{\partial x_1^{s_1}\ldots\partial x_n^{s_n}}\right|^p dv_P < \infty,$$

$$s_1 + \ldots + s_n = \bar{r} + s, \quad s = 1, 2, \ldots, k - \bar{r} - 1, \quad \eta \leqslant t_0.$$

6°. *There exists a neighborhood* O *of* $K^{(m)}$ *such that*

$$\bar{O}' \setminus \bar{K}^{(m)} \subset O \setminus \bar{K}^{(m)} \text{ and } F \equiv 0 \text{ on } E^n \setminus O'.$$

REMARK. If the manifold $K^{(m)}$ is infinitely differentiable, the function F can be chosen so as to be infinitely continuously differentiable on $E^n \setminus \bar{K}^{(m)}$. The theorem remains valid in the case when not one manifold $K^{(m)}$ has been given but a set of such manifolds separated from each other by a positive distance.

In the case when $K^{(m)}$ is a bounded and closed manifold the existence of functions satisfying conditions 2°– 4° has been proved by Nikol'skiĭ ([4], page 108).

Let us prove Theorem [8.3]. As usual, we assume that a system $\mathfrak{S} = \{V\}$ of canonical neighborhoods forming a locally finite covering of $K^{(m)}$ has been fixed. It can be assumed without loss of generality that every neighborhood $V \subset \mathfrak{S}$ is such that $V \subset O$ while the intersection $V \cap (\overline{K}^{(m)} \setminus K^{(m)})$ is empty ($\overline{K}^{(m)} \setminus K^{(m)}$ is a closed set). By virtue of the local finiteness of the covering $\mathfrak{S} = \{V\}$ it is possible to inscribe in it another locally finite covering $\mathfrak{S}^* = \{U\}$ of $K^{(m)}$, also consisting of canonical neighborhoods and such that for each $U \in \mathfrak{S}^*$ there exists a $V \in \mathfrak{S}$ for which $\overline{U} \subset V$ (see, for example, [33]).

Suppose $P \in K^{(m)}$. Then there exists a canonical neighborhood $U_P \in \mathfrak{S}^*$ such that $P \in U_P$. Consider the set of all possible balls Q_P, Q'_P, Q''_P with centers at P and radii $r_P, r_P/2, r_P/3$ respectively for which $\overline{Q}_P \subset U_P$. We now choose a countable system of points $P_i \in K^{(m)}$ and radii $r_i = r_{P_i}, i = 1, \cdots, n$, such that, if we put $Q_{P_i} = Q_i, Q'_{P_i} = Q'_i$ and $Q''_{P_i} = Q''_i$, the following conditions hold:

1°. For any point $P \in G = E^n \setminus (\overline{K}^{(m)} \setminus K^{(m)})$ there exists a neighborhood of it which intersects only a finite number of balls Q_i (and hence a finite number of balls Q'_i and Q''_i).

2°. The system $\{Q''_i\}$ forms a locally finite covering of $K^{(m)}$.

3°. $\overline{Q}_i \subset U^{(i)} \in \mathfrak{S}^*$ for all $i = 1, 2, \cdots$.

Let us establish the possibility of such a choice of points P_i and radii r_i. Since $K^{(m)}$ is an absolute F_σ set, it has the representation $K^{(m)} = \bigcup_1^\infty A_s$, where the A_s are compact sets, $A_s \subset A_{s+1}, s = 1, 2, \cdots$. For each point $P \in A_1$ we choose a bounded neighborhood W_P of it so that the intersection $\overline{W}_P \cap (\overline{K}^{(m)} \setminus K^{(m)})$ is empty. From the system $\{W_P\}_{P \in A_1}$ we select a finite covering and denote the intersection of $K^{(m)}$ with the union of its members by B_1. Clearly, the intersection $\overline{B}_1 \cap (\overline{K}^{(m)} \setminus K^{(m)})$ is empty. Suppose there have been defined open (in $K^{(m)}$) sets $B_1, \cdots, B_{\nu_0}$ such that $B_1 \subset B_2 \subset \cdots \subset B_{\nu_0} \subset K^{(m)}, A_\nu \subset B_\nu, \overline{B}_\nu$ is compact and the intersection $\overline{B}_\nu \cap (\overline{K}^{(m)} \setminus K^{(m)})$ is empty, $\nu = 1, \cdots, \nu_0$. Consider the compact set $\overline{B}_{\nu_0} \cup A_{\nu_0+1}$. For it, as for A_1, there exists an open (in $K^{(m)}$) set $\overline{B}_{\nu_0+1} \supset \overline{B}_{\nu_0} \cup A_{\nu_0+1}$ such that $\overline{B}_{\nu_0+1}$ is compact while the intersection $\overline{B}_{\nu_0+1} \cap (\overline{K}^{(m)} \setminus K^{(m)})$ is empty.

Thus $B_1 \subset B_2 \subset \cdots \subset B_\nu \subset \cdots \subset K^{(m)}$ and $\bigcup_\nu B_\nu = K^{(m)}$.

We put $C_1 = \overline{B}_1$ and $C_\nu = \overline{B}_\nu \setminus B_{\nu-1}, \nu = 2, 3, \cdots$. Then the C_ν are compact sets ($\nu = 1, 2, \cdots$) and $\bigcup_\nu C_\nu = K^{(m)}$. From the above-mentioned system of all possible balls Q''_P we select the subsystem consisting of those balls for which $P \in C_\nu$ and $r < 1/\nu$. This subsystem forms a covering of the compact set C_ν. We select from it a finite subcovering and denote it by $\mathfrak{S}'' = \{Q''_P\}$. Corresponding finite sets of the balls Q_P and Q'_P are denoted by $\mathfrak{S}_\nu$ and $\mathfrak{S}'_\nu$, while the balls themselves of the system $\mathfrak{S}_\nu$, $\mathfrak{S}'_\nu$ and $\mathfrak{S}''_\nu$ will be called balls of rank ν. We next arbitrarily number the centers P of all of the balls contained in at least one of the systems $\mathfrak{S}_\nu, \nu = 1, 2, \cdots$. Let us show that the resulting set of points P_i together with the corresponding balls from the systems $\mathfrak{S}_\nu, \mathfrak{S}'_\nu$ and $\mathfrak{S}''_\nu$ is the desired one.

Suppose $P \in G$, so that $\rho_P = \rho(P, \overline{K}^{(m)} \setminus K^{(m)}) > 0$. Then the ball $Q(P, \rho_P/2)$ with center at P and radius $\rho_P/2$ intersects only a finite number of the C_ν. Otherwise $\overline{Q}(P, \rho_P/2)$ would contain a limit point for an infinite sequence of compact sets C_ν that clearly can only belong to the set $\overline{K}^{(m)} \setminus K^{(m)}$, which is impossible. Therefore suppose the ball $Q(P, \rho_P/2)$ does not intersect any C_ν for $\nu \geqslant \nu_1$. We choose $\nu_2 \geqslant \nu_1$ such that $1/\nu_2 \leqslant \rho_P/12$. Then the ball $Q(P, \rho_P/3)$ does not intersect any balls of rank $\nu > \nu_2$, i.e. it intersects only a finite number of the selected balls Q_i.

From the fact that the balls of the system $\mathfrak{S}''_\nu$ cover the compact set $C_\nu, \nu = 1, 2, \cdots$, and the fact that $K^{(m)} = \bigcup_\nu C_\nu$ it follows that the set of balls $Q_i, i = 1, 2, \cdots$, forms a covering of $K^{(m)}$, while its local finiteness follows from the already proved condition 1°. Thus condition 2° is also satisfied, and as to condition 3°, it follows from the original choice of the balls Q. The existence of the needed system $\{Q_i\}$ is proved.

For every $U^{(i)} \in \mathfrak{S}^*$ there exists a $V \in \mathfrak{S}$ such that $\overline{U}^{(i)} \subset V$. Let $V^{(i)}$ denote one such V and let $V^{(i)}_n$ denote its preimage under the canonical mapping $P = x(u), P \in V^{(i)}, u \in V^{(i)}_u$. Thus with each point P_i there is associated a pair of canonical neighborhoods $U^{(i)}$ and $V^{(i)}$, either one of which may also be associated with other points P_i ($i = 1, 2, \cdots$).

According to the definition of the φ_s extendability of the system of functions $\{\psi_{\lambda^{(l)}_{m+1} \cdots \lambda^{(l)}_n}\}$ there exist functions $\tilde{f}_i(u)$ defined respectively on the $V^{(i)}_u$ that together with all of their derivatives are continuous everywhere on the $V^{(i)}_u$ except possibly the sets $\overset{\circ}{V}{}^{(i)}_u$ (see (8.6)). In addition,

$$\tilde{f}_i \in H^{(r)}_{p(n)}(M_i, V^{(i)}_u), \quad M_i \leqslant c^{(i)}_1 \sum M_{\lambda^{(l)}_{m+1} \cdots \lambda^{(l)}_n} + c^{(i)}_2 \sum \|\psi_{\lambda^{(l)}_{m+1} \cdots \lambda^{(l)}_n}\|^{(m)}_p, \tag{8.21}$$

$$\|\tilde{f}_i\|_{p\ V_u^{(i)}}^{(n)} \leqslant M, \tag{8.22}$$

$$\left.\frac{\partial^l \tilde{f}_i}{\partial u_{m+1}^{\lambda_{m+1}^{(l)}} \cdots \partial u_n^{\lambda_n^{(l)}}}\right|_{\overset{\circ}{V}{}_u^{(i)}} = \psi_{\lambda_{m+1}^{(l)} \cdots \lambda_n^{(l)}}[x(u)], \quad u \in \overset{\circ}{V}{}_u^{(i)}, \tag{8.23}$$

$$\int \cdots \int_{V_u^{(i)}} \varphi_s(r) \left| \frac{\partial^{\bar{r}+s} \tilde{f}(u)}{\partial u_1^{s_1} \cdots \partial u_n^{s_n}} \right|^p du_1 \ldots du_n < \infty, \tag{8.24}$$

$$s = 1, 2, \ldots, k - \bar{r} - 1.$$

We put $f_i(P) = \tilde{f}(x(u))$, where $P = x(u)$. Then by virtue of a theorem of Nikol'skiĭ on the behavior of functions belonging to H-classes under a change of the independent variables ([4], page 59) we get

$$f_i \in H_{p(n)}^{(r)}(M_i^*, U^{(i)}), \quad M_i^* \leqslant c_1^{\prime(i)} M_i + c_2^{\prime(i)} \|\tilde{f}_i\|_{p\ V_u^{(i)}}^{(n)} \tag{8.25}$$

and

$$\|f\|_{p\ V^{(i)}}^{(n)} \leqslant M_i^*. \tag{8.26}$$

The indicated theorem of Nikol'skiĭ can be applied as a result of the fact that the canonical mapping $P = u(x)$ is $k-1$ times continuously differentiable, $\bar{r} + 2 \leqslant k$, and the absolute value of the Jacobian is bounded from below by a positive constant. It also follows from this according to (8.24) that

$$\int \cdots \int_{V^{(i)}} \varphi_s(r(P)) \left| \frac{\partial^{\bar{r}+s} f_i(P)}{\partial x_1^{s_1} \cdots \partial x_n^{s_n}} \right|^p dx_1 \ldots dx_n < \infty, \quad s = 1, 2, \ldots, k - \bar{r} - 1, \tag{8.27}$$

$$P = (x_1, \ldots, x_n)$$

and that the functions f_i are $k-1$ times continuously differentiable on $V^{(i)} \setminus K^{(m)}$. Finally, according to definition (8.20) and equality (8.23) we have

$$\left.\frac{\partial^l f_i}{\partial N_{m+1}^{\lambda_{m+1}^{(l)}} \cdots \partial N_n^{\lambda_n^{(l)}}}\right|_{K^{(m)} \cap V^{(i)}} = \psi_{\lambda_{m+1}^{(l)} \cdots \lambda_n^{(l)}}. \tag{8.28}$$

For the subsequent proof of the theorem we make use of an idea of a construction developed by Nikol'skiĭ for extending functions from bounded and closed manifolds onto the whole space E^n without weight ([4], page 108), the origins of which go back to the works of Whitney and Hestenes [30, 31].

We first define a family of functions $F_i(P), i = 1, 2, \cdots$, as follows:

$$F_i(P) = \begin{cases} f_i(P) & \text{for } P \in Q_i, \\ 0 & \text{for } P \in E^n \setminus Q_i. \end{cases} \tag{8.29}$$

We next take a system of functions $\varphi_i(P), i = 1, 2, \cdots$, that are defined on E^n and satisfy the following conditions:

1°. The functions $\varphi_i(P)$ together with their derivatives up to order $k-1$ inclusively are continuous.

2°. $\varphi_i(P) = 0$ for $\rho(P, P_i) \leqslant r_i/3$.

3°. $\varphi_i(P) = 1$ for $\rho(P, P_i) \geqslant r_i/2$.

4°. The normal derivatives (mixed and not mixed) up to order $k-1$ inclusively of the functions φ_i are equal to zero on $K^{(m)}$.

The existence of such functions φ_i has been proved by Nikol'skiĭ ([4], page 115) for the case of bounded and closed manifolds. The proof of existence is completely analogously in our case, owing to the fact that it is connected only with the local properties of the manifold, and we will therefore not present it.

Let $\Phi_i = \varphi_1 \cdots \varphi_{i-1}(1-\varphi_i)$, $i = 1, 2, \cdots$. Then $|\Phi_i| \leqslant 1$ and $\Phi_i(P) = 0$ if $P \notin Q_i'$. This implies that the function $F_i\Phi_i$ is $k-1$ times continuously differentiable on E^n.

We define the desired function F by the equality

$$F = \sum_{i=1}^{\infty} F_i\Phi_i \tag{8.30}$$

and show that this definition is meaningful and that the resulting function satisfies conditions 1°–6° of the theorem being proved.

If $P \in \overline{K}^{(m)} \setminus K^{(m)}$, there does not exist a ball $Q_i \ni P$. In fact, $Q_i \subset V^{(i)} \in \mathfrak{S}$, while by assumption the intersection $V \cap (\overline{K}^{(m)} \setminus K^{(m)})$ is empty for every canonical neighborhood $V \in \mathfrak{S}$. Therefore $\varphi_i(P) = 1$, which implies $\Phi_i(P) = 0$ for all $i = 1, 2, \cdots$, i.e. $F(P) = 0$.

If $P \in G = E^n \setminus (\overline{K}^{(m)} \setminus K^{(m)})$, there exists a neighborhood $O(P)$ of P that intersects only a finite number of the balls Q_i. There therefore exists an i_0 such that $\varphi_i(P) = 1$ and hence $\Phi_i(P) = 0$ for $i \geqslant i_0$. It follows that $F(P) = \Sigma_1^{i_0} F_i(P)\Phi_i(P)$. If, in addition, $P \in G \setminus \overline{K}^{(m)}$, the neighborhood $O(P)$ can be chosen so that it does not intersect $\overline{K}^{(m)}$, and then $F = \Sigma_1^{i_0} F_i \Phi_i$ is a $k-1$ times continuously differentiable function on $O(P)$ since each summand $F_i\Phi_i$ is such a function. Thus (8.30) is meaningful and, what is more, we have shown that the function F satisfies condition 1° of Theorem [8.3].

We put $O' = \bigcup_i Q_i'$. Then $O' \subset O$ since $Q_i' \subset V^{(i)} \subset O$ and $\overline{O}' \setminus K^{(m)} \subset O \setminus \overline{K}^{(m)}$.

If $P \in E^n \setminus Q_i'$, then $\varphi_i(P) = 1$, and hence $\Phi_i(P) = 0$ for all $i = 1, 2, \cdots$, which implies $F(P) = 0$ for $P \in O'$. Requirement 6° is proved.

Now let $\mathfrak{G}$ be a domain such that $\overline{\mathfrak{G}}$ is compact and $\overline{\mathfrak{G}} \subset G$, and let $\mathfrak{G}_K = \mathfrak{G} \cap K^{(m)}$.

From the fact that $\overline{\mathfrak{G}}_K$ is also compact it follows that it intersects only a finite number of the balls Q_i. Suppose the intersection $Q_i \cap \overline{\mathfrak{G}}_K$ is empty for $i > i_0$. Then $W = \bigcup_1^{i_0} V^{(i)}$ forms a neighborhood of the compact set $\overline{\mathfrak{G}}_K$. There therefore exists an $\eta > 0$ such that the cylindroid $K_\eta^{(n)}(\mathfrak{G}_K) \subset W$. Hence for $s = 1, 2, \cdots, k-r-1$, using (8.29) and (8.30), we get

$$\int_{K_\eta^{(n)}(\mathfrak{G}_K)} \cdots \int \varphi_s(r) \left| \frac{\partial^{\overline{r+s}} F}{\partial x_1^{s_1} \ldots \partial x_n^{s_n}} \right|^p dx_1 \ldots dx_n \leqslant \sum_{i=1}^{i_0} \int_{V^{(i)}} \cdots \int \varphi_s(r) \left| \frac{\partial^{\overline{r+s}} F}{\partial x_1^{s_1} \ldots \partial x_n^{s_n}} \right|^p dx_1 \ldots dx_n$$

$$\leqslant c \sum_{i=1}^{i_0} \int_{V^{(i)}} \cdots \int \varphi_s(r) \left| \frac{\partial^{\overline{r+s}} f}{\partial x_1^{s_1} \ldots \partial x_n^{s_n}} \right|^p dx_1 \ldots dx_n,$$

where $c > 0$ is a constant. It follows by virtue of (8.27) that condition 5° is satisfied.

The compact set $\overline{\mathfrak{G}}$ intersects only a finite number of the balls Q_i. Otherwise the intersection $\overline{\mathfrak{G}} \cap (\overline{K}^{(m)} \setminus K^{(m)})$ would not be empty. Suppose the intersection $\overline{\mathfrak{G}} \cap Q_i$ is empty for $i > i_0$. Then

$$F = \sum_{i=1}^{i_0} F_i \Phi_i$$

on $\mathfrak{G}$, and, since F_i satisfies conditions (8.25), (8.26) and (8.28) on Q_i while Φ_i is such that $F_i\Phi_i$ is a smoothing of F_i to a function that identically vanishes outside Q_i', the Φ_i having continuous derivatives up to order $k-1$ inclusively that are uniformly bounded in absolute value on E^n, it follows that the function F satisfies conditions 2° and 3° on G. The proof of this fact as well as the proof of condition 4°, which by definition has a local character, is an insignificant modification of the proof of the above mentioned theorem of Nikol′skiĭ ([4], pages 111–115) for the case of bounded and closed manifolds $K^{(m)}$. We will therefore not reproduce it here.

We note a special case of Theorem [8.3], which we will essentially make use of in the sequel.

[8.4]. *Suppose* $1 \leqslant p < \infty$, $K^{(n-1)}$ *is an* $(n-1)$*-dimensional manifold of order of smoothness* $k \geqslant 2$, O *is a neighborhood of it,* $\rho > 0$, $\bar{\rho} = \rho + \beta$, $r = \rho + 1/p = \bar{r} + \alpha$ *and* $\bar{r} + 2 \leqslant k$. *Further, for each* $\lambda = 0, 1, \cdots, \bar{\rho}$ *suppose given a function* $\psi_\lambda \in H^{(\rho-\lambda)}_{p(n-1)}(M_\lambda, K^{(n-1)})$. *Finally, suppose* $G = E^n \setminus (\bar{K}^{(n-1)} \setminus K^{(n-1)})$, $\mathfrak{G}$ *is a domain in* E^n *such that* $\mathfrak{G}$ *is compact and* $\bar{\mathfrak{G}} \subset G$, $\mathfrak{G}_K = \mathfrak{G} \cap K^{(n-1)}$ *and* $r(P)$ *is the distance from a point* P *to* $K^{(n-1)}$ *"with respect to the normals" to* $K^{(n-1)}$. *Then there exists a function* $F(x_1, \cdots, x_n)$ *defined on* E^n *and satisfying the following requirements.*

1°. F *is continuously differentiable on* $E^n \setminus K^{(n-1)}$ *up to order* $k-1$ *inclusively.*

2°. $F \in H^{(r)}_{p(n)}(M, \mathfrak{G})$, *where*

$$M \leqslant c_1 \sum_{\lambda=0}^{\bar{\rho}} M_\lambda + c_2 \sum_{\lambda=0}^{\bar{\rho}} \| \psi_\lambda \|^{(n-1)}_{\substack{p \\ K^{(n-1)}}},$$

the constants c_1 *and* c_2 *not depending on* M_λ *or* ψ_λ.

3°. $\| F \|^{(n)}_{\substack{p \\ \mathfrak{G}}} \leqslant M.$

4°. $\partial^\lambda F / \partial N^\lambda \big/_{K^{(n-1)}} = \psi_\lambda$, $\lambda = 0, 1, \cdots, \bar{\rho}$.

5°. *There exists an* $\eta > 0$ *such that for any* $\epsilon > 0$

$$\int_{K^{(n)}_\eta(\mathfrak{G}_K)} \cdots \int r^{p(s-\alpha)+\epsilon} \left| \frac{\partial^{\bar{r}+s} F}{\partial x_1^{s_1} \ldots \partial x_n^{s_n}} \right|^p dx_1 \ldots dx_n < \infty.$$

6°. *There exists a neighborhood* O' *of* $K^{(n-1)}$ *such that*

$$\bar{O}' \setminus \bar{K}^{(n-1)} \subset O \setminus \bar{K}^{(n-1)} \quad \text{and} \quad F \equiv 0 \quad \text{on} \quad E^n \setminus O'.$$

In the case $p = \infty$ *it is only necessary to change the inequality in condition* 5° *to*

5$^\infty$.
$$\left\| r^{s-\alpha+\epsilon} \frac{\partial^{\bar{r}+s} F}{\partial x_1^{s_1} \ldots \partial x_n^{s_n}} \right\|^{(n)}_{\substack{\infty \\ K^{(n)}_\eta(\mathfrak{G}_K)}} < \infty.$$

This theorem is an immediate consequence of [8.2] and [8.3].

It will be shown below (see Theorem [12.3]) that in the sense of the order of growth of the derivatives (i.e. in the sense of the exponent of the distance r in condition 5°) this extension will be best to within an arbitrary $\epsilon > 0$.

Namely, we prove an imbedding theorem, the converse of the present theorem, from which it will follow that if a system of functions ψ_λ, $\lambda = 0, 1, \cdots, \rho$, of the sort discussed in Theorem [8.4] were to have an extension F whose derivatives of some order have a smaller order of growth under an approach to the manifold $K^{(n-1)}$ (i.e. if the exponent of the distance r were less than in assertion 5° or, respectively, in 5^∞), the original functions ψ_λ, $\lambda = 0, 1, \cdots, \bar{\rho}$, would belong to H-classes with higher exponents than is assumed in the conditions of Theorem [8.4].

Definition. Let $K^{(m)}$ be an m-dimensional manifold. We will say that a function f is of class $\bar{H}_p^{(r)}(M, K^{(m)})$ or, more precisely, of class $\bar{H}_p^{(r)}(M, K^{(m)}, K^{*(m)})$ if there exist an m-dimensional manifold $K^{*(m)}$ and a function f^* defined on $K^{*(m)}$ such that

1°. $K^{(m)}$ is a submanifold of $K^{*(m)}$,

2°. $f^* = f$ on $K^{(m)}$,

3°. $f^* \in H_p^{(r)}(M, K^{*(m)})$.

In the sequel we will say for the sake of simplicity that a set K is a *piece* of an m-dimensional manifold if there exists an m-dimensional manifold $K^{(m)}$ for which K is an m-dimensional submanifold. If the boundary of a domain G consists of a finite number of pieces of manifolds, we will always assume that each piece K is part of a manifold K^* such that $K^* \setminus \bar{K}$ lies in $E^n \setminus G$. Only such K^* will be considered in this connection.

Various criteria for $\bar{H}$-classes when $K^{*(m)}$ is a flat manifold can be found in a paper of Nikol′skiĭ [44]. These and similar criteria (see the author's paper [13]) can be generalized to "curved manifolds." One should first construct local extensions of the function in a neighborhood of each boundary point of the manifold on which the original function is given. This can be done by reducing the problem with the use of canonical mappings to the already analyzed flat case. These extensions are then smoothed out by a "Whitney" construction into a single function. A detailed presentation of this method can be found in the author's thesis [43] (see the proof of Theorem [8.5] there).

In §12 below we give a criterion for H-classes that is based on imbedding theorems with weight.

§9. Function extensions to a domain with a piecewise smooth boundary

In the present section we consider a particular problem concerning the extension of a function given on the boundary of a domain of a special form with a piecewise smooth boundary. The results obtained will be essentially used in the sequel.

Let $\overset{+}{E}{}^n$ denote the halfspace $x_n > 0$. It will be assumed that the domain $G \subset \overset{+}{E}{}^n$ and that its boundary K is representable in the form

$$K = \bigcup_{\kappa=1}^{\kappa_0} \overline{K}_\kappa \cup \bigcup_{\lambda=1}^{\lambda_0} L_\lambda \cup \bigcup_{\mu=1}^{\mu_0} M_\mu, \tag{9.1}$$

where each set K_κ is a finite piece of an $(n-1)$-dimensional manifold of order of smoothness $k \geqslant 2$, each L_λ is a domain in the hyperplane $E^{n-1} = [x_n = 0]$ and each M_μ is simply an $\overline{(n-1)}$-dimensional bounded and closed manifold of order of smoothness $k \geqslant 2$ lying in the halfspace $\overset{+}{E}{}^n$. Thus the domain G is generally multipily connected.

We will say that a piece K_κ does not intersect the hyperplane E^{n-1} at a zero angle if its outward normal $\overline{N}_P$ at every point $P \in \overline{K}_\kappa \cap E^{n-1}$ forms a positive angle α_{N_P} with the coordinate axis O_{x_n}. When this condition is satisfied,

$$0 < \theta_0 \leqslant \alpha_{N_P} \leqslant \pi, \quad P \in \overline{K}_\kappa \cap E^{(n-1)},$$

where θ_0 is a constant. If this is true for all $\kappa = 1, \cdots, \kappa_0$, we will simply say that the part

$$\overset{+}{K} = \bigcup_{\kappa=1}^{\kappa_0} K_\kappa \tag{9.2}$$

of the boundary does not intersect the hyperplane E^{n-1} at a zero angle.

For every function F which together with its first order weak derivatives is defined on G we put

$$D_\alpha(F) = \int_G \cdots \int x_n^\alpha \sum_{i=1}^n \left(\frac{\partial F}{\partial x_i}\right)^2 dx_1 \ldots dx_n. \tag{9.3}$$

Thus $D_0(F) = D(F)$ is the Dirichlet integral. Let us prove a lemma.

[9.1.1]. *Suppose that the boundary K of a finite domain $G \subset \overset{+}{E}{}^n$ satisfies the above conditions and that a part $\overset{+}{K}$ of it does not intersect the hyperplane $E^{n-1} = [x_n = 0]$ at a zero angle. Then there exist pairwise disjoint neighborhoods*

$U_K^{(\kappa)}, V_K^{(\kappa)}, U_L^{(\lambda)}, V_L^{(\lambda)}, U_M^{(\mu)}$ *and* $V_M^{(\mu)}$ *of the sets* K_κ, L_λ *and* M_μ,

$$\overline{U}_K^{(\kappa)} \setminus K \subset V_K^{(\kappa)} \setminus K;\ \overline{U}_L^{(\lambda)} \setminus K \subset V_L^{(\lambda)} \setminus K;\ \overline{U}_M^{(\mu)} \setminus K \subset V_M^{(\mu)} \setminus K$$

$(\kappa = 1, \cdots, \kappa_0; \lambda = 1, \cdots, \lambda_0; \mu = 1, \cdots, \mu_0)$ *respectively and a function* Φ *defined and bounded on* E^n *satisfying the following conditions*:

1°. Φ *is continuously differentiable on* $E^n \setminus (\overset{\mp}{K} \cap E^{n-1})$.

2°. $\Phi(P) = 1$ *for* $P \in \bigcup_{\kappa=1}^{\kappa_0} U_K^{(\kappa)} \cup \bigcup_{\lambda=1}^{\lambda_0} U_L^{(\lambda)} \cup \bigcup_{\mu=1}^{\mu_0} U_M^{(\mu)}$.

3°. $\Phi(P) = 0$ *for* $P \in E^n \setminus [\bigcup_{\kappa=1}^{\kappa_0} V_K^{(\kappa)} \cup \bigcup_{\lambda=1}^{\lambda_0} V_L^{(\lambda)} \cup \bigcup_{\mu=1}^{\mu_0} V_M^{(\mu)}]$.

4°. $D_\alpha(\Phi) < \infty$.

PROOF. We choose a pair K_κ and L_λ such that the intersection $\overline{K}_\kappa \cap \overline{L}_\lambda$ is not empty. Suppose for the sake of definiteness that K_1 and L_1 are such a pair and let $\gamma = \overline{K}_1 \cap \overline{L}_1$. We assume for the sake of simplicity that the closed domain $\overline{L}_1$ does not intersect any other piece $\overline{K}_\kappa$ $(\kappa \neq 1)$. If this is not the case, it will be necessary to carry out for each nonempty intersection $\overline{K}_\kappa \cap L_1$ the construction described below for γ.

The manifold piece K_1 is a submanifold of some manifold $\overline{K}_1^*$.[10] For each point $P_0 \in \gamma$ we choose a proper neighborhood V of it to which there corresponds the representation

$$x_i = x_i(u_1, \ldots, u_{n-1}),\ i = 1, 2, \ldots, n, \tag{9.4}$$

the rank of the matrix $\|\partial x_i/\partial u_j\|$ $(i = 1, \cdots, n; j = 1, \cdots, n-1)$ being equal to $n-1$. If the part $\overset{+}{K}$ of the boundary does not intersect the hyperplane $x_n = 0$ at P_0 at either a zero angle or an angle equal to π, at least one of the minors obtained from the above matrix by removing one of the first $n-1$ rows is different from zero at P_0. Suppose this is true, for example, for the minor obtained by removing the first row. Then there exists a neighborhood $V' \subset V$ of P_0 in which the manifold K_1^* has the representation

$$x_1 = f(x_2, \ldots, x_n), \tag{9.5}$$

and hence $\gamma' = \gamma \cap V'$ will be given by the representation

$$x_1 = f(x_2, \ldots, x_{n-1}, 0).$$

---

(10) See page 76.

Let $\bar{N}'_P = (\beta_1, \cdots, \beta_{n-1}, 0)$ denote the inward unit normal to γ' at a point $P \in \gamma'$ in the hyperplane $x_n = 0$. Then

$$\beta_1 = \pm \frac{1}{\sqrt{1 + \sum_{i=2}^{n-1} f_{x_i}^2}}, \quad \beta_j = \mp \frac{f_{x_j}}{\sqrt{1 + \sum_{i=2}^{n-1} f_{x_i}^2}}, \quad j = 2, \ldots, n-1.$$

Consider the mapping

$$\begin{aligned} x_i &= f(x_2, \ldots, x_{n-1}, 0) + \xi_1\beta_1, \\ x_i &= \xi_i, \quad i = 2, \ldots, n. \end{aligned} \tag{9.6}$$

Its Jacobian $\partial(x_1, \cdots, x_n)/\partial(\xi_1, \cdots, \xi_n) = \beta_1 \neq 0$ for $P \in \gamma'$, and hence there exists a neighborhood V'' of P_0 in which $\xi_1, \cdots, \xi_n$ can be regarded as local curvilinear coordinates. Equation (9.5) takes the form

$$F(\xi_1, \ldots, \xi_n) \equiv f(\xi_2, \ldots, \xi_{n-1}, 0) + \xi_1\beta_1 - f(\xi_2, \ldots, \xi_n) = 0.$$

From the fact that $\partial F/\partial \xi_1 = \beta_1 \neq 0$ it follows that there exists a neighborhood U of P_0 in which the representation of the manifold $K_1^* \cap U$ has the form

$$\xi_1 = \varphi(\xi_2, \ldots, \xi_n). \tag{9.7}$$

It can be assumed in this connection that the function φ has bounded derivatives in U and that $\bar{U}$ does not intersect any part of the boundary K of the domain G except K_1 and L_1.

Let $P = (x_1, \cdots, x_n) \in U \cap \gamma$ and let $E^2_{\xi_1\xi_n}$ be the plane passing through the point P that is parallel to the Ox_n axis and the vector $\bar{N}'_P$. Then the intersection of K_1^* with $E^2_{\xi_1\xi_n}$ has the representation

$$\xi_1 = \varphi_0(\xi_n) = \varphi(x_2, \ldots, x_{n-1}, \xi_n).$$

in U. The derivative $\partial\xi_1/\partial\xi_n$ is bounded and therefore, if $\tan \alpha(P) = d\varphi_0(\xi_1)/d\xi_n$, the function $\alpha(P)$ can be regarded as a continuous function of P such that there exists a constant θ_P for which

$$|\alpha(P)| < \frac{\pi}{2} - \theta_{P_0}, \quad P \in U. \tag{9.8}$$

We use ρ, θ to denote polar coordinates in the plane $E^2_{\xi_1\xi_2} = E^2_{\xi_1\xi_2}(P)$ and put

$$C[P, r, \Delta] = \{(\rho, \theta): \rho < r, \theta \in \Delta\}, \tag{9.9}$$

where Δ is a certain interval. Let $K_1^*(P) = K_1^* \cap V'' \cap E^2_{\xi_1 \xi_n}(P)$. It follows from (9.8) that there exist constants $r_{P_0} > 0, \theta'_{P_0} > 0$ and a neighborhood $W \subset U$ of P_0 such that

$$K_1^* \cap C[P, r_{P_0}, [0, 2\pi)] \subset C[P, r_{P_0}, (\theta'_{P_0}, \pi)], \ P \in \gamma \cap W.$$

And if the angle formed at P_0 by the normal $\overline{N}_P$ to K_1^* with the Ox_n axis is equal to π, we have the analogous inclusion

$$K_1^* \cap C[P, r_{P_0}, [0, 2\pi)] \subset C\left[P, r_{P_0}, \left(\theta'_{P_0}, \frac{4}{3}\pi\right)\right], \ P \in \gamma \cap W, \theta'_{P_0} = \frac{\pi}{2},$$

for which in place of (9.5) one should make use of the analogously obtained representation $x_n = g(x_1, \cdots, x_{n-1})$ of K_1^* in some neighborhood of P_0.

We now take a finite covering $\{W_j\}$ of the set γ by neighborhoods of the indicated type corresponding to the points $P_1, \cdots, P_{j_0}$, and we denote by r_0 and θ_0 respectively the least of the numbers r_{P_j} and $\theta'_{P_j}, j = 1, \cdots, j_0$.

Thus for any $P \in \gamma$

$$K_1^* \cap C[P, r_0, [0, 2\pi)] \subset C\left[P, r_0, \left(\theta_0, \frac{4}{3}\pi\right)\right], \ \theta_0 > 0.$$

We put

$$C[r, \Delta] = \bigcup_{P \in \gamma} C[P, r, \Delta]. \tag{9.10}$$

Then, if $\Delta = [0, 2\pi)$, the union $C[r, \Delta]$ is a neighborhood of γ in E^n. We now choose an $\eta > 0$ such that the cylindroids $K_\eta^{(n)}(K_1)$ and $K_\eta^{(n)}(L_1)$ (see (8.8)) satisfy the following conditions: if

$$A_K = K_\eta^{(n)}(K_1) \setminus \overline{C}\left[\frac{r_0}{2}, [0, 2\pi)\right],$$

$$A_L = K_\eta^{(n)}(L_1) \setminus \overline{C}\left[\frac{r_0}{2}, [0, 2\pi)\right],$$

then

$$\begin{aligned} A_K \cap C[r_0, [0, 2\pi)] &\subset C\left[r_0, \left(\frac{5}{6}\theta_0, \frac{3}{2}\pi\right)\right], \\ A_L \cap C[r_0, [0, 2\pi)] &\subset C\left[r_0, \left(-\frac{\theta_0}{6}, \frac{\theta_0}{6}\right)\right]. \end{aligned} \tag{9.11}$$

In addition, we will assume that the cylindroids $K_\eta^{(n)}(K_1)$ and $K_\eta^{(n)}(L_1)$ are separated by a positive distance from the other parts $K_2, \cdots, K_{\kappa_0}, L_2, \cdots, L_{\lambda_0}$ and $M_1, \cdots, M_{\mu_0}$ of the boundary.

Let

$$V_K^{(1)} = A_K \cup C\left[r_0, \left(\frac{1}{2}\theta_0, \frac{3}{2}\pi\right)\right],$$

$$V_L^{(1)} = A_L \cup C\left[r_0, \left(-\frac{\theta_0}{2}, \frac{\theta_0}{2}\right)\right].$$

$V_K^{(1)}$ and $V_L^{(1)}$ are neighborhoods of K_1 and L_1 that are separated by a positive distance from the other parts of the boundary.

Let $\psi(t)$ be an infinitely differentiable function on the real line such that $\psi(t) = 0$ for $t \leqslant 0$ and $\psi(t) = 1$ for $t \geqslant 1$. We define a function $\Psi(P)$ on the set $V = V_K^{(1)} \cup V_L^{(1)} \cup C[r_0, [0, 2\pi)]$ in the following manner. If $P \in V \setminus C[r_0, [0, 2\pi)]$, we put $\Psi(P) = 1$. If $P \in C[r_0, [0, 2\pi)]$, there exists a point $P_0 \in \gamma$ such that $P \in C[P_0, r_0, [0, 2\pi)]$. Denoting by ρ, θ the polar coordinates of P in the above mentioned plane $E^2(P_0)$, we now put

$$\Psi(P) = \begin{cases} 1 & \text{if } -\frac{\theta_0}{6} < \theta < \frac{\theta_0}{6} \text{ or if } \frac{2}{3}\theta_0 < \theta < \frac{7}{6}\pi, \\ 0 & \text{if } \rho = 0, \text{ if } \frac{1}{3}\theta_0 < \theta < \frac{2}{3}\theta_0, \\ & \text{or if } \frac{4}{3}\theta_0 < \theta < 2\pi - \frac{\theta_0}{3}, \\ \psi\left(2 - \frac{6\theta}{\theta_0}\right) & \text{if } \frac{\theta_0}{6} \leqslant \theta \leqslant \frac{1}{3}\theta_0, \\ \psi\left(\frac{6\theta}{\theta_0} - 4\right) & \text{if } \frac{2}{3}\theta_0 \leqslant \theta \leqslant \frac{5}{6}\theta_0, \\ \psi\left(4 - \frac{3\theta}{\pi}\right) & \text{if } \frac{7}{6}\pi \leqslant \theta \leqslant \frac{4}{3}\pi. \end{cases} \tag{9.12}$$

The constructed function $\Psi(P)$ is equal to unity in sectors containing points of the boundary K of the domain G and is equal to zero in sectors separating K_1 and L_1. Further, it is infinitely continuously differentiable in $C[P_0, r_0, [0, 2\pi)]$ for $\rho \neq 0$. Therefore $\Psi(P)$ is a continuously differentiable function of $x_1, \cdots, x_n$ throughout the domain $V \setminus \gamma$ (see conditions (9.11)).

We consider another neighborhood of the set $K_1 \cup L_1$:

$$V_1 = (V_K^{(1)} \cup V_L^{(1)}) \cap \left\{K^{(n)}_{\frac{1}{2}\eta}(K_1) \cup K^{(n)}_{\frac{1}{2}\eta}(L_1) \cup C\left[\frac{1}{2}r_0, [0, 2\pi)\right]\right\},$$

$\bar{V}_1$ is compact and $\bar{V}_1 \subset V$. Therefore the function Ψ can be extended by the usual methods (see, for example, Theorem [6.1]) from V_1 to a function Φ_{11} defined on all of E^n, continuously differentiable on $E^n \setminus \gamma$, coinciding with Ψ on V_1 and equal to zero outside V.

We put

$$U_K^{(1)} = \left\{K^{(n)}_{\frac{1}{2}-\eta}(K_1) \setminus C\left[\frac{1}{4}r_0, [0, 2\pi)\right]\right\} \cup C\left[\frac{1}{2}r_0, \left(\frac{5}{6}\theta_0, \frac{7}{6}\pi\right)\right],$$

$$U_L^{(1)} = \left\{K^{(n)}_{\frac{1}{2}-\eta}(L_1) \setminus C\left[\frac{1}{4}r_0, [0, 2\pi)\right]\right\} \cup C\left[\frac{1}{2}r_0, \left(-\frac{1}{6}\theta_0, \frac{1}{6}\theta_0\right)\right].$$

Then

$$\Phi_{11}(P) = 1 \quad \text{for} \quad P \in U_K^{(1)} \cup U_L^{(1)} \text{ and } \Phi_{11}(P) = 0 \quad \text{for} \quad P \in E^n \setminus (V_K^{(1)} \cup V_L^{(1)}).$$

Let us estimate the integral $D_\alpha(\Phi_{11})$. To this end we introduce the notation

$$B = E^n \setminus C\,[r_0, [0, 2\pi)]$$

and divide the integral under consideration into two integrals:

$$D_\alpha(\Phi_{11}) = I' + I'',$$

where

$$I' = \int_B \cdots \int x_n^\alpha \sum_1^n \left(\frac{\partial \Phi_{11}}{\partial x_i}\right)^2 dx_1, \ldots, dx_n < \infty,$$

since the function Φ_{11} is continuously differentiable on the closed domain B and is equal to zero outside a ball of sufficiently large radius.

For each point $P = (x_1, \cdots, x_n) \in \gamma$ we choose a neighborhood U of it which in local coordinates $\xi_1, \cdots, \xi_n$ of type (9.6) or in some other similar coordinates has the form $|\xi_i - x_i| < \epsilon$, $i = 2, \cdots, n-1$, $\xi_1^2 + \xi_n^2 < r_0^2$, with $\xi_n = x_n$. We take a covering of γ by such neighborhoods $U_1, \cdots, U_{\nu_0}$ corresponding to some points $(x_1^{(\nu)}, \cdots, x_n^{(\nu)}) \in \gamma$ and constants $\epsilon_\nu > 0$, $\nu = 1, 2, \cdots, \nu_0$. Then

$$I'' = \int_{C[r_0, [0, 2\pi)]} \cdots \int |x_n^\alpha| \sum_{i=1}^n \left(\frac{\partial \Phi_{11}}{\partial x_i}\right)^2 dx_1 \ldots dx_n \leqslant \sum_{\nu=1}^{\nu_0} \int_{U_\nu} \cdots \int |x_n^\alpha| \sum_{i=1}^n \left(\frac{\partial \Phi_{11}}{\partial x_i}\right)^2 dx_1 \ldots dx_n.$$

By virtue of the fact that the transformation of coordinates from $(x_1, \cdots, x_n)$ to $(\xi_1, \cdots, \xi_n)$ can be assumed without loss of generality to be boundedly continuously differentiable with a Jacobian whose absolute value is bounded from below by a positive constant, it suffices to prove the finiteness of the integral

$$I_\nu = \int_{U_\nu} \cdots \int |\xi_n^\alpha| \sum_{i=1}^{n} \left(\frac{\partial \Phi_{11}}{\partial \xi_i}\right)^2 d\xi_1 \ldots d\xi_n.$$

We now go over in the plane of the variables ξ_1 and ξ_n to polar coordinates ρ, θ. It then follows from formula (9.12) that $\partial\Phi_{11}/\partial\xi_i = 0$, $i = 2, \cdots, n-1$, $\partial\Phi_{11}/\partial\rho = 0$, and that there exists a constant $c > 0$ for which $|\partial\Phi_{11}/\partial\theta| < c$. Therefore, noting that

$$\left(\frac{\partial \Phi_{11}}{\partial \xi_1}\right)^2 + \left(\frac{\partial \Phi_{11}}{\partial \xi_n}\right)^2 = \left(\frac{\partial \Phi_{11}}{\partial \rho}\right)^2 + \frac{1}{\rho^2}\left(\frac{\partial \Phi_{11}}{\partial \theta}\right)^2,$$

we get

$$I_\nu = \int_{U_\nu} \cdots \int |\rho \sin\theta|^\alpha \left(\frac{1}{\rho}\frac{\partial \Phi_{11}}{\partial \theta}\right)^2 \rho \, d\rho \, d\theta \, d\xi_2 \ldots d\xi_{n-1}$$

$$\leqslant C \int_{x_2-\varepsilon_\nu}^{x_2+\varepsilon_\nu} d\xi_2 \ldots \int_{x_{n-1}-\varepsilon_\nu}^{x_{n-1}+\varepsilon_\nu} d\xi_{n-1} \int_0^{2\pi} d\theta \int_0^{r_0} \frac{d\rho}{\rho^{1-\alpha}} < \infty \quad \text{for } \alpha > 0.$$

Hence $I'' < \infty$ and therefore

$$D_\alpha(\Phi_{11}) < \infty.$$

To complete the proof of the lemma one should use the method described above to construct for each of the remaining domains L_λ of the hyperplane E^{n-1}, $\lambda = 2, \cdots, \lambda_0$, and for the corresponding K_{κ_j} ($j = 1, \cdots, j_\lambda$ and κ_j takes values from the numbers $2, \cdots, \kappa_0$) such that the intersection $\overline{K}_{\kappa_j} \cap \overline{L}_\lambda$ is not empty a function $\Phi_{\lambda\kappa_1\cdots\kappa_{j_\lambda}}$ defined on E^n, continuously differentiable on the domain $E^n \setminus \bigcup_j \overline{K}_{\kappa_j} \cap L_\lambda$, equal to zero outside the set $V_L^{(\lambda)} \cup V_K^{(\kappa_1)} \cup \cdots \cup V_K^{(\kappa_{j_\lambda})}$, where $V_L^{(\lambda)}, V_K^{(\kappa_1)}, \cdots, V_K^{(\kappa_{j_\lambda})}$ are pairwise disjoint neighborhoods of the sets $L_\lambda, K_{\kappa_1}, \cdots, K_{\kappa_\lambda}$ respectively that are separated by a positive distance from the remaining parts of the boundary, and such that $D_\alpha(\Phi_{\lambda\kappa_1\cdots\kappa_{j_\lambda}}) < \infty$. In addition, there must exist neighborhoods $U_K^{(\kappa_j)} \subset V_K^{(\kappa_j)}$, $j = 1, \cdots, j_\lambda$, and $U_L^{(\lambda)} \subset V_L^{(\lambda)}$ of the sets K_{κ_j} and L_λ respectively on which $\Phi_{\lambda\kappa_1\cdots\kappa_{j_\lambda}}$ is equal to 1. All of this can be effected by virtue of the construction mentioned above.

Finally, for each manifold M_μ one should take ϵ and $\epsilon/2$ neighborhoods $V_M^{(\mu)}$ and $U_M^{(\mu)}$, put $\Psi_\mu(P) = 1$ for $P \in V_M^{(\mu)}$ and extend Ψ_μ by means of a Whitney construction, for example, from the neighborhood $U_M^{(\mu)}$ onto all of E^n

as a continuously differentiable function $\Phi_\mu(P)$ such that $\Phi_\mu(P) = \Psi_\mu(P)$ for $P \in U_M^{(\mu)}$ and $\Phi_\mu(P) = 0$ outside $V_M^{(\mu)}$. Clearly, for Φ_μ we have

$$D_\alpha(\Phi_\mu) < \infty, \quad \mu = 1, 2, \ldots, \mu_0.$$

All of the considered neighborhoods $V_M^{(\mu)}$, $V_L^{(\lambda)}$ and $V_K^{(\kappa)}$ can be chosen so as to be pairwise disjoint. The desired function Φ is therefore determined as a sum of the constructed Φ-functions with various indices:

$$\Phi = \Sigma\Phi_{\lambda\kappa_1 \ldots \kappa_{j_\lambda}} + \Sigma\Phi_\mu.$$

[9.1]. *Suppose the boundary* K *of a finite domain* $G \subset \overset{+}{E}{}^n$, $n > 1$, *satisfies the above conditions and the part* $\overset{+}{K}$ *of it does not intersect the hyperplane* E^{n-1} *at a zero angle. Further, suppose given on the boundary* K *a function* f *or, more explicitly, on the pieces* K_κ *functions* f_{K_κ}, $\kappa = 1, 2, \cdots, \kappa_0$ *(see* (9.1)), *on the pieces* L_λ *functions* f_{L_λ}, $\lambda = 1, 2, \cdots, \lambda_0$, *and on the manifolds* M_μ *functions* f_{M_μ}, $\mu = 1, 2, \cdots, \mu_0$, *such that*

$$f_{K_\kappa} \in \overline{H}{}^{\rho_{K_\kappa}}_{2(n-1)}(K_\kappa), \quad \rho_{K_\kappa} > \frac{n-1}{2}, \quad \kappa = 1, 2, \ldots, \kappa_0,$$

$$f_{L_\lambda} \in \overline{H}{}^{\rho_{L_\lambda}}_{2(n-1)}(L_\lambda), \quad \rho_{L_\lambda} > \frac{n-1}{2}, \quad \lambda = 1, 2, \ldots, \lambda_0,$$

$$f_{M_\mu} \in H^{\rho_{M_\mu}}_{2(n-1)}(M_\mu), \quad \rho_{M_\mu} > \frac{1}{2}, \quad \mu = 1, 2, \ldots, \mu_0.$$

Then there exists a function F *defined and continuously differentiable on* G *and having the properties*

1°. $F|_{K_\kappa} = f_{K_\kappa}$ $(\kappa = 1, \cdots, \kappa_0)$; $F|_{L_\lambda} = f_{L_\lambda}$ $(\lambda = 1, \cdots, \lambda_0)$; $F|_{M_\mu} = f_{M_\mu}$ $(\mu = 1, \cdots, \mu_0)$;

2°. $D_\alpha(F) < \infty$.

PROOF. By virtue of the conditions $f_{K_\kappa} \in \overline{H}{}^{\rho_{K_\kappa}}_{2(n-1)}(K_\kappa)$, $f_{L_\lambda} \in \overline{H}{}^{\rho_{L_\lambda}}_{p(n-1)}(L_\lambda)$ and the definition of the $\overline{H}$-classes (see §8) there exist $(n-1)$-dimensional manifolds K_κ^* and L_λ^* in E^n for which $\overline{K}_\kappa \subset K_\kappa^*$, $\overline{L}_\lambda \subset L_\lambda^*$, $K_\kappa^* \setminus \overline{K}_\kappa \subset E^n \setminus \overline{G}$ and $L_\lambda^* \setminus \overline{L}_\lambda \subset E^n \setminus \overline{G}$, and the functions $f_{K_\kappa}^*$ and $f_{L_\lambda}^*$ defined respectively on K_κ^* and L_λ^* such that $f_{K_\kappa}^* = f_{K_\kappa}$ on K_κ, $f_{L_\lambda}^* = f_{L_\lambda}$ on L_λ, $f_{K_\kappa}^* \in H^{\rho_{K_\kappa}}_{2(n-1)}(K_\kappa^*)$ and $f_{L_\lambda}^* \in$

$H_{2(n-1)}^{\rho_{L_\lambda}}(L_\lambda^*)$. We next choose bounded neighborhoods $\mathfrak{G}_{K_\kappa}$ and $\mathfrak{G}_{L_\lambda}$ of the sets $\bar{K}_\kappa$ and $\bar{L}_\lambda$ in such a way that $\mathfrak{G}_{K_\kappa} \supset \bar{G}, \mathfrak{G}_{L_\lambda} \supset \bar{G}, \bar{\mathfrak{G}}_{K_\kappa} \subset E^n \setminus (\bar{K}_\kappa^* \setminus K_\kappa^*)$ and $\bar{\mathfrak{G}}_{L_\lambda} \subset E^n \setminus (\bar{L}_\lambda^* \setminus L_\lambda^*)$. According to Theorem [8.4] the functions $f_{K_\kappa}^*$ and $f_{L_\lambda}^*$ can be extended onto all of E^n as functions $F_{K_\kappa}^*$ and $F_{L_\lambda}^*$ respectively that are continuously differentiable on $E^n \setminus \bar{K}_\kappa^*$ and $E^n \setminus \bar{L}_\lambda^*$, with

$$F_{K_\kappa}^* \in H_{2(n)}^{\rho_{K_\kappa}+\frac{1}{2}}(\mathfrak{G}_{K_\kappa}), \quad F_{L_\lambda}^* \in H_{2(n)}^{\rho_{L_\lambda}+\frac{1}{2}}(\mathfrak{G}_{L_\lambda}), \tag{9.13}$$

$$F_{K_\kappa}^*\big|_{K_\kappa^*} = f_{K_\kappa}^*, \quad F_{L_\lambda}^*\big|_{L_\lambda^*} = f_{L_\lambda}^*, \; \kappa = 1, 2, \ldots, \kappa_0, \; \lambda = 1, 2, \ldots, \lambda_0. \tag{9.14}$$

The conditions of the theorem imply that $\rho_{K_\kappa} + ½ > 1$ and $\rho_{L_\lambda} + ½ > 1$. It therefore follows from (9.13) that $F_{K_\kappa}^*$ and $F_{L_\lambda}^*$ have derivatives with summable square in $\mathfrak{G}_{K_\kappa}$ and $\mathfrak{G}_{L_\lambda}$, and hence by virtue of the choice of $\mathfrak{G}_{K_\kappa}$ and $\mathfrak{G}_{L_\lambda}$

$$\int_G \cdots \int \sum_{i=1}^n \left(\frac{\partial F_{K_\kappa}^*}{\partial x_i}\right)^2 dx_1 \ldots dx_n < \infty, \quad \int_G \cdots \int \sum_{i=1}^n \left(\frac{\partial F_{L_\lambda}^*}{\partial x_i}\right)^2 dx_1 \ldots dx_n < \infty. \tag{9.15}$$

Further, $\rho_{K_\kappa} + ½ > n/2$ and $\rho_{L_\lambda} + ½ > n/2$. Therefore, extending according to Theorem [6.1], for example, the functions $F_{K_\kappa}^*$ and $F_{L_\lambda}^*$ from some domains $\mathfrak{G}'_{K_\kappa}$ and $\mathfrak{G}'_{L_\lambda}$ ($\bar{\mathfrak{G}} \subset \mathfrak{G}'_{K_\kappa} \subset \bar{G}'_{K_\kappa} \subset \mathfrak{G}_{K_\kappa}$ and analogously for $\mathfrak{G}'_{L_\lambda}$) onto all of E^n with preservation of the H-class in a sense, and then applying an imbedding theorem of Nikol'skiĭ ([3], Theorem 12, page 26),[11] we get that

$$F_{K_\kappa}^* \in H_{\infty(n)}^{\alpha_{K_\kappa}}(G), \; \alpha_{K_\kappa} > 0, \quad F_{L_\lambda}^* \in H_{\infty(n)}^{\alpha_{L_\lambda}}(G), \; \alpha_{L_\lambda} > 0,$$

which implies the existence of a constant $A > 0$ such that

$$|F_{K_\kappa}^*| \leqslant A, \quad |F_{L_\lambda}^*| \leqslant A \text{ on } G, \; \kappa = 1, 2, \ldots, \kappa_0, \; \lambda = 1, 2, \ldots, \lambda_0. \tag{9.16}$$

As to the functions f_{M_μ}, according to the same Theorem [8.4] they can be extended at once as functions F_{M_μ} defined on all of E^n, continuously differentiable on $E^n \setminus M_\mu$ and satisfying the conditions

(11) This theorem asserts in particular that if a function $f \in H_{p(n)}^r$ then $f \in H_{\infty(n)}^\rho$ for $\rho = r - n/p > 0$.

$$F_{M_\mu} \in H_{2(n)}^{\rho_{M_\mu}+\frac{1}{2}}, \tag{9.17}$$

$$F_{M_\mu}|_{M_\mu} = f_{M_\mu} \quad (\mu = 1, 2, \ldots, \mu_0). \tag{9.18}$$

From condition (9.17) and the fact that $\rho_{M_\mu} + ½ > 1$ we have in particular

$$\int_G \cdots \int \sum_{i=1}^{n} \left(\frac{\partial F_{M_\mu}}{\partial x_i} \right)^2 dx_1 \ldots dx_n < \infty. \tag{9.19}$$

Suppose now Φ is the function defined in Lemma [9.1.1]. We put

$$\Phi_{K_\kappa} = \begin{cases} \Phi \text{ on } V_K^{(\kappa)} \\ 0 \text{ on } G \setminus V_K^{(\kappa)} \end{cases}, \quad \Phi_{L_\lambda} = \begin{cases} \Phi \text{ on } V_L^{(\lambda)} \\ 0 \text{ on } G \setminus V_L^{(\lambda)} \end{cases}; \quad \Phi_{M_\mu} = \begin{cases} \Phi \text{ on } V_M^{(\mu)} \\ 0 \text{ on } G \setminus V_M^{(\mu)} \end{cases}$$

(for the notation see the conditions of the lemma). From the indicated lemma it follows that all of the functions obtained are continuously differentiable on G. We define the desired function F by the formula

$$F = \sum_{\kappa=1}^{\kappa_0} F_{K_\kappa}^* \Phi_{K_\kappa} + \sum_{\lambda=1}^{\lambda_0} F_{L_\lambda}^* \Phi_{L_\lambda} + \sum_{\mu=1}^{\mu_0} F_{M_\mu} \Phi_{M_\mu}. \tag{9.20}$$

It is not difficult to see that F is continuously differentiable on G. In addition,

$$\begin{gathered} F|_{K_\kappa} = f_{K_\kappa} \ (\kappa = 1, 2, \ldots, \kappa_0), \ F|_{L_\lambda} = f_{L_\lambda} \ (\lambda = 1, 2, \ldots, \lambda_0), \\ F|_{M_\mu} = f_{M_\mu} (\mu = 1, 2, \ldots, \mu_0). \end{gathered} \tag{9.21}$$

In fact, in each of the neighborhoods $U_K^{(\kappa)}$, $U_L^{(\lambda)}$ and $U_M^{(\mu)}$ (see the conditions of the lemma) F coincides with $F_{K_\kappa}^*$, $F_{L_\lambda}^*$ or $F_{M_\mu}^*$ respectively, since according to the lemma $\Phi_{K_\kappa} = 1$, $\Phi_{L_\lambda} = 1$ and $\Phi_{M_\mu} = 1$ in these neighborhoods respectively; inasmuch as condition (9.21) is local, it follows from (9.14) and (9.18).

It remains to prove condition 1° of the theorem. It suffices to do this for each summand of (9.20).

Consider $F_{K_1}^* \Phi_{K_1}$, for example:

$$\int_G \cdots \int x_n^\alpha \sum_{i=1}^n \left[\frac{\partial (F_{K_1}^* \Phi_{K_1})}{\partial x_i}\right]^2 dx_1 \ldots dx_n = \int_{v_K^{(1)} \cap G} \cdots \int x_n^\alpha \sum_{i=1}^n \left[\frac{\partial (F_{K_1}^* \Phi_{K_1})}{\partial x_i}\right]^2 dx_1 \ldots dx_n$$

$$= \int_{V_K^{(1)} \cap G} \cdots \int x_n^\alpha F_{K_1}^{*2} \sum_{i=1}^n \left(\frac{\partial \Phi_{K_1}}{\partial x_i}\right)^2 dx_1 \ldots dx_n + \int_{V_K^{(1)} \cap G} \cdots \int x_n^\alpha \Phi_{K_1}^2 \sum_{i=1}^n \left[\frac{\partial F_{K_1}^*}{\partial x_i}\right]^2 dx_1 \ldots dx_n$$

$$= J_1 + J_2.$$

For the first integral, using estimate (9.16) and Lemma [9.1.1], we have

$$J_1 \leqslant A \int_{V_K^{(1)} \cap G} \cdots \int x_n^\alpha \sum_{i=1}^n \left(\frac{\partial \Phi_{K_1}}{\partial x_i}\right)^2 dx_1 \ldots dx_n < \infty.$$

For the second integral, noting that by virtue of the same lemma there exists a constant $A' > 0$ such that $|\Phi_{K_1}| < A'$ on G while by virtue of the boundedness of G there exists a constant $a > 0$ for which $G \subset \{(x_i): 0 < x_n < a\}$ and applying (9.15), we get

$$J_2 \leqslant A'^2 \int_{V_K^{(1)} \cap G} \cdots \int x_n^\alpha \sum_{i=1}^n \left(\frac{\partial F_{K_1}^*}{\partial x_i}\right)^2 dx_1 \ldots dx_n \leqslant A'^2 a^\alpha \int_G \cdots \int \sum_{i=1}^n \left(\frac{\partial F_{K_1}^*}{\partial x_i}\right)^2 dx_1 \ldots dx_n < \infty.$$

The finiteness of the corresponding integrals for the other summands of (9.20) is proved analogously. The theorem is proved.

In concluding this section we note that the result can without doubt be significantly sharpened at the expense of a more perfect method of extending the boundary functions. In order to establish an exact estimate of the class of boundary functions admitting an extension with a finite integral $D_\alpha(F)$ one should first study the properties of the boundary values, of the function for which $D_\alpha(F) < \infty$, on the part $\overset{+}{K}$ of the boundary K of the domain G under an approach to the hyperplane $x_n = 0$. Some necessary conditions for this can be found in a paper of M. I. Višik [23].

For the sake of domains with corners the question of function extensions in the "absence of a degeneracy", in particular for $\alpha = 0$, has been fully studied to a sufficient extent in the works of Nikol'skiĭ [34, 35].

§10. The variation of boundaries principle

In the present section we consider a direct application of our theorems which is connected with the solution of boundary problems for partial differential equation by the variation of boundaries method. It consists in the following. Suppose given (i) a domain G with a boundary K on which is prescribed a function f, and (ii) a corresponding equation $L(u) = 0$, it being required to find a solution u of this equation that takes the value f on K. The variation of boundaries method permits one to replace a boundary problem for a given function and a given domain by the same boundary problem but generally for a smoother function and a different domain with a subsequent limit passage. To apply it one should extend the function f as a function F that is continuously differentiable a sufficient number of times according to Theorem [8.4], for example, and consider a sequence of domains G_ν, $\nu = 1, 2, \cdots$, that have sufficiently smooth boundaries K_ν and contract to G in such a way that $G_1 \supset G_2 \supset \cdots \supset G_\nu \supset \cdots \supset G$. One must then solve the boundary problem for the equation $L(u) = 0$ in the domain G_ν under the boundary conditions $u|_{K_\nu} = F|_{K_\nu}$. We denote the solution by u_ν. For certain classes of domains and types of equations it can be expected that the sequence of functions u_ν will converge in G to a solution of the original boundary problem in G as $\nu \to \infty$.

If the boundary problem in G_ν is solved by the ordinary variational method, the above method will be called the dual variational method. This method is discussed by us on several occasions, but a detailed study of it is not a goal of the present monograph, the main goal of which in the part concerning applications is a solution by the classical variational method of ellipitc equations degenerate on the boundary of the domain.

We will not solve any new problems, but we illustrate the dual variational method by considering as an example the well-studied Dirichlet problem for the Laplace operator. A heuristic argument speaking in favor of this method in regard to that problem is the stability of the solution of the Dirichlet problem under an approximation of the domain from without, which has been studied by Keldyš and Lavrent′ev (see, for example, [1]). It will be shown that the class of admissible boundary functions for the generalized variational method is significantly wider than for the ordinary variational method, which is connected, of course, with the appearance of an additional limit passage.

We will essentially use a result of N. M. Gjunter. Following the terminology introduced by him, a surface K bounding a finite domain G in three-dimensional space will be called a surface of $\mathcal{J}_r(B, \lambda)$, $r = 0, 1, \cdots, 0 < \lambda < 1$, if it is a Ljapunov surface and has the property that, for any point $P \in K$ and any orthogonal cartesian coordinate system (ξ, η, ζ) such that the $O\xi$ and $O\eta$ axes lie in the tangent plane to K at P while the $O\zeta$ axis lies along the normal $\overline{N}_P$ to K at P, there exists a neighborhood of P in which K has a representation $\zeta = \zeta(\xi, \eta) \in H_\infty^{(r+\lambda)}(B)$.

GJUNTER'S THEOREM. *Suppose a surface K bounds a finite domain G, $K \in Л_{r+1}(B, \lambda)$, $r \geqslant 0$, and suppose given on K a function $f \in H_\infty^{(l+\lambda)}(M)$, $0 \leqslant l \leqslant r+1$. Then the harmonic function which is the solution of the Dirichlet problem for G is of class $H_\infty^{(l+\lambda')}(M')$, where λ' is any number satisfying the condition $0 < \lambda' < \lambda$ while $M' \leqslant aM$, the constant a depending only on K and the choice of λ' but not on f* ([36], *Chapter* V, §3).

Without going into a detailed analysis of the constant $a > 0$ we note only that in the case when the surface K has an order of smoothness $k \geqslant 2$ there exists an $\epsilon > 0$ such that a single constant $a > 0$ can be chosen for all of the boundaries K_η, $0 < \eta < \epsilon$, of the η-neighborhoods G_η of G.

It will always be assumed in what follows in this section that the surface K satisfies all of the above conditions. We prove the following theorem.

[10.1]. *Suppose given on the boundary K of a finite domain $G \subset E^3$ a function $f \in H_\infty^{(\rho)}$, $0 < \rho < 1$, and suppose F is an extension of it onto E^3 (see [8.4]) satisfying the following conditions.*

1°. *$F = f$ on K.*

2°. *F is continuously differentiable on $E \setminus K$.*

3°. *$F \in H_\infty^{(\rho)}(M)$.*

4°. *$\eta^{1-\rho+\epsilon}\|\partial F/\partial x_i\|_\infty < c$, $i = 1, 2, 3$, the constant c not depending on η.*

Further, suppose $\eta_\nu > 0$, $\lim_{\nu\to\infty}\eta_\nu = 0$ and u_ν is a harmonic function defined on G_{η_ν} and taking the value $f_\nu = F|_{K_{\eta_\nu}}$ on K_{η_ν}.

Then the sequence u_ν converges uniformly on G to a harmonic function u such that $u|_K = f$, with

$$|u - u_\nu| < c'\eta_\nu^\beta,$$

where $\beta < \rho - \epsilon$, $\epsilon > 0$ and the constant c' does not depend on η.

PROOF. By virtue of the fact that F is continuously differentiable on $E^n \setminus K$ the function $f_\nu = F|_{K_{\eta_\nu}}$ is of class $H_\infty^{(1)}(M_\nu, K_{\eta_\nu})$ (and even in the strong sense, i.e. in the sense of first differences; see §6), where

$$M_\nu \leqslant b \sum_{i=1}^{3} \left\| \frac{\partial F}{\partial x_i} \right\|_{\substack{\infty \\ K_{\eta_\nu}}}, \tag{10.1}$$

in which the constant $b > 0$ can be chosen so that it does not depend on ν. Then for a harmonic function u_ν we will have according to Gjunter's Theorem

$$u_\nu \in H_\infty^{(\lambda)}(M', G_\nu), \tag{10.2}$$

where λ is chosen so that $1 - \rho + \epsilon < \lambda < 1$ while $M' \leqslant aM$ (a does not depend on ν). Suppose next that points P and P_ν lie on a normal to K and $P \in K_\nu$, $P_\nu \in K_{\eta_\nu}$. Then by virtue of (10.2), (10.1) and condition 4° of the theorem we get

$$|u_\nu(P) - u_\nu(P_\nu)| \leqslant M'\eta_\nu^\lambda \leqslant ab \sum_{i=1}^{3} \left\| \frac{\partial F}{\partial x_i} \right\|_\infty \eta_\nu^\lambda \leqslant 3abc\eta_\nu^\beta, \quad \beta = \lambda + \rho - 1 - \varepsilon. \tag{10.3}$$

But from condition 3° of the theorem it follows that

$$|F(P_\nu) - F(P)| < M\eta_\nu^\rho, \tag{10.4}$$

in which $F(P_\nu) = u_\nu(P)$ and $F(P) = f(P)$. Therefore (10.3) and (10.4) imply

$$|u_\nu(P) - f(P)| < 3abc\eta_\nu^\beta + M\eta_\nu^\rho \leqslant c'\eta_\nu^\beta, \tag{10.5}$$

where $c' > 0$ is a constant. It follows for any $\mu = 1, 2, \cdots$ that $u_{\nu+\mu}(P) - u_\nu(P) \to 0$ as $\nu \to \infty$, which means by virtue of the maximum principle that the sequence u_ν converges uniformly on $\bar{G}$ to a harmonic function u clearly satisfying by virtue of (10.5) the conditions $u|_K = f$ and

$$|u_\nu(P) - u(P)| \leqslant c'\eta_\nu^\beta.$$

Thus not only have we established the uniform convergence of the sequence of functions $u_\nu(P)$, but we have also determined its "rate of convergence" (cf. [1]).

The above theorem shows that the Dirichlet problem can be solved by the dual variational method in the case when the boundary function f is of class $H_\infty^{(\rho)}$ on K and $\rho > 0$, since the function F always exists (see Theorem [8.4]) while a harmonic function u_ν can be found that minimizes the Dirichlet integral over G_ν under the corresponding boundary conditions. Theorem [10.1] together with its proof can be carried over in a natural way to the n-dimensional case.

Let us show that the generalized variational method can sometimes give a solution when the ordinary variational method does not apply. We consider the classical *Hadamard's example* (see, for example, [2], page 97). Let

$$\varphi(\theta)=\sum_{n=1}^{\infty}\frac{\cos n^4\theta}{n^2},\quad 0\leqslant\theta\leqslant 2\pi.$$

As is well known, the function

$$u=u(\rho,\theta)=\sum_{k=1}^{\infty}\frac{\cos n^4\theta}{n^2}\rho^{n^4}$$

is harmonic in the interior of the unit disk, continuous on its closure and satisfies the boundary condition $u(1,\theta)=\varphi(\theta)$, while the Dirichlet integral $D(u)=\infty$. Thus in the present case the variational principle is not suitable for solving the Dirichlet problem.

Let $T_n(\theta)=\Sigma_{k=1}^{n}\cos k^4\theta/k^2$ and $R(\theta)=\Sigma_{k=n+1}^{\infty}\cos k^4\theta/k^2$. Clearly, $T_n(\theta)$ is a trigonometric polynomial of degree $\nu_n=n^4$ and there exists a constant c such that $|R_n(\theta)|\leqslant c/n$. Therefore

$$\max_{0\leqslant\theta\leqslant 2\pi}|\varphi(\theta)-T_n(\theta)|\leqslant\frac{c}{n}=\frac{c}{\nu_n^{1/4}}.$$

The sequence ν_n contains geometric progressions, for example, $(2^4)^k$, $k=1,2,\cdots$. It follows from a theorem of S. N. Bernšteĭn ([37], *Works* page 29) that $\varphi\in H_\infty^{(1/4)}$. But this means according to Theorem [10.1] that the Dirichlet problem in the case of Hadamard's example is solvable by the dual variational method.

In conclusion we note that, besides the indicated generalized variational method involving a variation of the boundary of the domain and an extension of the boundary values, there is another natural approach, viz. the one having a *variation of the boundary values* under a fixed boundary, which is based on an approximation of the given boundary function by means of smoother functions. Here one can take advantage of the fact that any function belonging to L_2 on the boundary can be approximated with any degree of accuracy in the sense of convergence in the mean by arbitrarily smootn functions (assuming, of course, a corresponding smoothness in the boundary). This method requires a discussion of the connection between the convergence in the mean of the boundary values and the convergence of the corresponding solutions on the interior of the domain.

SECOND CHAPTER

WEIGHTED IMBEDDING THEOREMS

§11. Weighted imbedding theorems for function classes on flat manifolds

In the present section we adopt the following notation and assumptions: $E^m = \{(x_i) \in E^n : x_{m+1} = \cdots = x_n = 0\}$, $1 \leqslant m < n$, $1/p + 1/q = 1$, $1 \leqslant p \leqslant \infty$, $r = \sqrt{x_{m+1}^2 + \cdots + x_n^2}$, $0 < \eta < a < \infty$, $\xi - a - \eta$, Γ is a finite or infinite domain in the hyperplane E^m, $\Gamma_\eta = \{P : \rho(P, E^m \setminus \Gamma) > \eta\}$, $\Gamma_a^{(n)} = K_a^{(n)}(\Gamma)$, $\Gamma_{\eta\xi}^{(n)} = K_\xi^{(n)}(\Gamma_\eta)$ (see (8.8)), $\Gamma_{+aj}^{(n)} = K_{+aj}^{(n)}(\Gamma)$, $\Gamma_{+\xi j}^{(n)} = K_{+\xi j}^{(n)}(\Gamma)$, $\Gamma_{\eta(+\xi)j}^{(n)} = K_{+\xi j}^{(n)}(\Gamma_\eta)$, $\Gamma_{-aj}^{(n)} = K_{-aj}^{(n)}(\Gamma)$, $\Gamma_{\eta(-\xi)j}^{(n)} = K_{-\xi j}^{(n)}(\Gamma_\eta)$ (see (8.12)), $\Gamma_{+aj}^{(n-1)} = K_{+aj}^{(n-1)}(\Gamma)$, $\Gamma_{-aj}^{(n-1)} = K_{-aj}^{(n-1)}(\Gamma)$, $\Gamma_{+\xi j}^{(n-1)} = K_{+\xi j}^{(n-1)}(\Gamma)$ (see (8.9) and (8.10)), $j = m + 1, \cdots, n$. The case $\Gamma = E^m$ is not excluded; here one takes $\Gamma_\eta = E^n$. If $m = n - 1$, the index j is dropped.

In each of the following theorems it is assumed that a function f together with any weak derivatives of it appearing in the statement of the theorem is defined in a cylindroid $\Gamma_a^{(n)}$. It is further assumed that for the $j = m + 1, \cdots, n$ under consideration and for almost all $(x_1, \cdots, x_{j-1}, 0, x_{j+1}, \cdots, x_n)$ the function f has a limit for $x_j \to a - 0$ $(x_j \to -a + 0)$ which is taken as the boundary value of f on $\Gamma_{+aj}^{(n-1)}$ (on $\Gamma_{-aj}^{(n-1)}$). An analysis of this assumption is carried out in [11.6]. For the sake of simplicity and uniformity of the estimates we will take $a \geqslant 1$. The notation for a weighted norm is used (see (5.1) and (6.2)). In Theorems [11.1]–[11.7] we assume that $p < \infty$.

[11.1]. *Suppose* j *is fixed* $(j = m + 1, \cdots, n)$, $\epsilon > 0$ *and* $\alpha \geqslant p - 1$. *Then there exist constants* c_1 *and* c_2 *not depending on* f *or* ϵ *such that*

$$\|f\|^{(n)}_{p(r,\alpha-p+\epsilon),\ \Gamma^{(n)}_{+aj}} \leqslant \frac{1}{(\alpha+1-p+\epsilon)^{\frac{1}{p}}}\left[c_1 \|f\|^{(n-1)}_{p,\ \Gamma^{(n-1)}_{+aj}} + \frac{c_2}{\epsilon^{\frac{1}{q}}} \|f_{x_j}\|^{(n)}_{p(r,\alpha),\ \Gamma^{(n)}_{+aj}}\right].^{(1)} \qquad (11.1)$$

---

(1) At the present time a similar result for $m = n - 1$ and $\epsilon = 0$ has been obtained by S. V. Uspenskiĭ.

PROOF. Suppose that the sake of definiteness that $j = n$. Then

$$\|f\|^{(n)}_{p(r,\alpha-p+\varepsilon)} = \left\{\int_{\Gamma}\cdots\int dx_1\ldots dx_m \int_{-a}^{a} dx_{m+1}\ldots \int_0^a r^{\alpha-p+\varepsilon}|f(x_1,\ldots,x_n)|^p dx_n\right\}^{\frac{1}{p}}$$
$$\Gamma^{(n)}_{+aj}$$

$$= \left\{\int_{\Gamma}\cdots\int\int_{-a}^{a}\cdots\int_{-a}^{a}\int_0^a r^{\alpha-p+\varepsilon}\,|f(x_1,\ldots,x_{n-1},a)\right.$$

$$\left. - \int_{x_n}^{a} f_{x_n}(x_1,\ldots,x_{n-1},t)\,dt\,|^p\,dx_1\ldots dx_n\right\}^{\frac{1}{p}} \tag{11.2}$$

$$\leqslant \left\{\int_{\Gamma}\cdots\int\int_{-a}^{a}\cdots\int_{-a}^{a}\int_0^a r^{\alpha-p+\varepsilon}|f(x_1,\ldots,x_{n-1},a)|^p\,dx_1\ldots dx_n\right\}^{\frac{1}{p}}$$

$$+\left\{\int_{\Gamma}\cdots\int\int_{-a}^{a}\cdots\int_{-a}^{a}\int_0^a r^{\alpha-p+\varepsilon}\left|\int f_{x_n}(x_1,\ldots,x_{n-1},t)\,dt\right|^p dx_1\ldots dx_n\right\}^{\frac{1}{p}}$$

$$= I_1 + I_2.$$

If $\alpha \geqslant p$, we have

$$r^{\alpha-p+\varepsilon} \leqslant (n-m)^{\frac{\alpha-p+\varepsilon}{2}} a^{\alpha-p+\varepsilon}$$

Therefore

$$I_1 = \left\{\int_{\Gamma}\cdots\int\int_{-a}^{a}\cdots\int_{-a}^{a}\int_0^a r^{\alpha-p+\varepsilon}|f(x_1,\ldots,x_{n-1},a)|^p\,dx_1\ldots dx_n\right\}^{\frac{1}{p}}$$

$$\leqslant (n-m)^{\frac{\alpha-p+\varepsilon}{2p}} a^{\frac{\alpha+\varepsilon}{p}-1}$$

(10:3)

$$\times\left\{\int_0^a dx_n \int_{\Gamma}\cdots\int\int_{-a}^{a}\cdots\int_{-a}^{a}|f(x_1,\ldots,x_{n-1},a)|^p\,dx_1\ldots dx_{n-1}\right\}^{\frac{1}{p}}$$

$$= (n-m)^{\frac{\alpha-p+\varepsilon}{2p}} a^{\frac{1+\alpha+\varepsilon}{p}-1}\|f\|^{(n-1)}_{p,\,\Gamma^{(n-1)}_{+an}} \leqslant a_1\|f\|^{(n-1)}_{p,\,\Gamma^{(n-1)}_{+an}},$$

where

$$a_1 = (n-m)^{\frac{\alpha-p+1}{2p}} a^{\frac{2+\alpha-p}{p}-1}$$

does not depend on ϵ $(\epsilon \leqslant 1)$.

If $p-1 \leqslant \alpha < p$, the fact that $0 < x_n < r$ implies

$$r^{\alpha-p+\varepsilon} < \frac{1}{x_n^{p-\alpha-\varepsilon}}.$$

Therefore

$$I_1 \leqslant \left\{ \int_0^a \frac{dx_n}{x_n^{p-\alpha-\epsilon}} \int \ldots \int_{\Gamma} \int_{-a}^{a} \ldots \int_{-a}^{a} |f(x_1, \ldots, x_{n-1}, a)|^p \, dx_1 \ldots dx_{n-1} \right\}^{\frac{1}{p}}$$

$$\leqslant \|f\|_{p,\,\Gamma^{(n-1)}_{+an}}^{(n-1)} \left\{ \int_0^a \frac{dx_n}{x_n^{p-\alpha-\varepsilon}} \right\}^{\frac{1}{p}} \leqslant \frac{a_2}{(\alpha-p+1+\varepsilon)^{\frac{1}{p}}} \|f\|_{p,\,\Gamma^{(n-1)}_{+an}}^{(n-1)},$$

where $a_2 = a^{(\alpha+2)/p-1}$ does not depend on ϵ $(\epsilon \leqslant 1)$. From these inequalities it follows that there exists a constant $c_1 > 0$ not depending on ϵ such that in both cases

$$I_1 \leqslant \frac{c_1}{(\alpha-p+1+\varepsilon)^{\frac{1}{p}}} \|f\|_{p,\,\Gamma^{(n-1)}_{+an}}^{(n-1)}. \tag{11.3}$$

We now estimate the integral I_2. Applying the generalized Minkowski inequality to the following expression in brackets, we get

$$I_2 = \left\{ \int \ldots \int_{\Gamma} \int_{-a}^{a} \ldots \int_{-a}^{a} \left[\int_0^a r^{\alpha-p+\varepsilon} \right.\right.$$
$$\left.\left. \times \left| \int_{x_n}^{a} f_{x_n}(x_1, \ldots, x_{n-1}, t)\, dt \right|^p dx_n \right] dx_1 \ldots dx_{n-1} \right\}^{\frac{1}{p}}$$
$$\leqslant \left\{ \int \ldots \int_{\Gamma} \int_{-a}^{a} \ldots \int_{-a}^{a} \left[\int_0^a \left(\int_0^t r^{\alpha-p+\varepsilon} \right.\right.\right. \tag{11.4}$$
$$\left.\left.\left. \times |f_{x_n}(x_1, \ldots, x_{n-1}, t)|^p \, dx_n \right)^{\frac{1}{p}} dt \right]^p dx_1 \ldots dx_{n-1} \right\}^{\frac{1}{p}}$$
$$= \left\{ \int \ldots \int_{\Gamma} \int_{a}^{a} \ldots \int_{-a}^{a} \left[\int_0^a |f_{x_n}(x_1, \ldots, x_{n-1}, t)| \right.\right.$$
$$\left.\left. \times \left(\int_0^t r^{\alpha-p+\varepsilon} \, dx_n \right)^{\frac{1}{p}} dt \right]^p dx_1 \ldots dx_{n-1} \right\}^{\frac{1}{p}}.$$

We again consider two cases. Suppose first $\alpha \geqslant p$ and

$$r_{(t)} = \sqrt{x_{m+1}^2 + \ldots + x_{n-1}^2 + t^2},$$

Then for $0 < x_n < t$ we have $r < r_{(t)}$. Therefore

$$\left(\int_0^t r^{\alpha - p + \varepsilon} dx_n \right)^{\frac{1}{p}} < r_{(t)}^{\frac{\alpha - p + \varepsilon}{p}} t^{\frac{1}{p}} \leqslant r_{(t)}^{\frac{\alpha + \varepsilon}{p} - 1 + \frac{1}{p}} = r_{(t)}^{\frac{\alpha}{p} - \frac{1}{q} + \frac{\varepsilon}{p}}.$$

Applying now Hölder's inequality to the expression in brackets in (11.4), we get

$$\left[\int_0^a | f_{x_n}(x_1, \ldots, x_{n-1}, t) | \left(\int_0^t r^{\alpha - p + \varepsilon} dx_n \right)^{\frac{1}{p}} dt \right]^p$$

$$\leqslant \left[\int_0^a r_{(t)}^{\frac{\alpha}{p} - \frac{1}{q} + \frac{\varepsilon}{p}} | f_{x_n}(x_1, \ldots, x_{n-1}, t) | \, dt \right]^p$$

$$\leqslant \left(\int_0^a \frac{dt}{r_{(t)}^{1 - \frac{q\varepsilon}{p}}} \right)^{\frac{p}{q}} \int_0^a r_{(t)}^{\alpha} | f_{x_n}(x_1, \ldots, x_{n-1}, t) |^p \, dt$$

$$\leqslant \left(\int_0^a \frac{dt}{t^{1 - \frac{q\varepsilon}{p}}} \right)^{\frac{p}{q}} \int_0^a r_t^{\alpha} | f_{x_n} |^p \, dt \leqslant \frac{b}{\varepsilon^{\frac{p}{q}}} \int_0^a r_{(t)}^{\alpha} | f_{x_n} |^p \, dt,$$

where the constant $b = (p/q)^{p/q} a^{p/q}$ does not depend on ϵ $(\epsilon \leqslant 1)$.

If on the other hand $p - 1 \leqslant \alpha < p$, by proceeding in an analogous manner we get (for $\epsilon < p - \alpha$)

$$\left[\int_0^a | f_{x_n}(x_1, \ldots, x_{n-1}, t) | \left(\int_0^t r^{\alpha - p + \varepsilon} \, dx_n \right)^{\frac{1}{p}} dt \right]^p$$

$$\leqslant \left[\int_0^a | f_{x_n}(x_1, \ldots, x_{n-1}, t) | \left(\int_0^t \frac{dx_n}{x_n^{p - \alpha - \varepsilon}} \right)^{\frac{1}{p}} dt \right]^p$$

$$= \frac{1}{\alpha + 1 - p + \varepsilon} \left[\int_0^a | f_{x_n} | \, t^{\frac{\alpha + 1 + \varepsilon - p}{p}} \, dt \right]^p$$

$$\leqslant \frac{1}{\alpha + 1 - p + \varepsilon} \left[\int_0^a | f_{x_n} | \, r^{\frac{\alpha + 1 + \varepsilon - p}{p}} \, dt \right]^p =$$

$$= \frac{1}{\alpha + 1 - p + \varepsilon}\left[\int_0^a r^{\frac{\alpha}{p}} \, |f_{x_n}| \frac{dt}{r^{\frac{1}{q} - \frac{\varepsilon}{p}}}\right]^p$$

$$\leqslant \frac{1}{\alpha + 1 - p + \varepsilon}\left(\int_0^a \frac{dt}{r_{(t)}^{1 - \frac{q\varepsilon}{p}}}\right)^{\frac{p}{q}} \int_0^a r_{(t)}^{\alpha} \, |f_{x_n}|^p \, dt$$

$$\leqslant \frac{b}{(\alpha + 1 - p + \varepsilon)\,\varepsilon^{\frac{p}{q}}} \int_0^a r_{(t)}^{\alpha} \, |f_{x_n}|^p \, dt.$$

Thus in both cases there exists a constant $c_2^p > 0$ not depending on ϵ $(\epsilon \leqslant 1)$ such that

$$\left[\int_0^a |f_{x_n}(x_1, \ldots, x_{n-1}, t)| \left(\int_0^t r^{\alpha - p + \varepsilon} \, dx_n\right)^{\frac{1}{p}} dt\right]^{\frac{1}{p}}$$

$$\leqslant \frac{c_2^p}{(\alpha + 1 - p + \varepsilon)\,\varepsilon^{\frac{p}{q}}} \int_0^a r_{(t)}^{\alpha} \, |f_{x_n}|^p \, dt.$$

Substituting this estimate into (11.4) and replacing the letter t by the letter x for the sake of simplicity, we get

$$I_2 \leqslant \frac{c_2}{(\alpha + 1 - p + \varepsilon)^{\frac{1}{p}} \varepsilon^{\frac{1}{q}}} \left\{ \int \ldots \int_{\Gamma} \int_{-a}^{a} \ldots \int_{-a}^{a} \int_0^a r^{\alpha} \right.$$

$$\left. \times |f_{x_n}(x_1, \ldots, x_n)|^p \, dx_1 \ldots dx_n \right\}^{\frac{1}{p}} = \frac{c_2}{(\alpha + 1 - p + \varepsilon)^{\frac{1}{p}} \varepsilon^{\frac{1}{q}}} \| f_{x_n} \|^{(n)}_{p\,(r, \alpha)\, \Gamma^{(n)}_{+an}}. \tag{11.5}$$

Inequalities (11.3) and (11.5) prove the desired inequality (11.1). In a completely analogous manner we can obtain the estimate

$$\| f \|^{(n)}_{p\,(r,\, \alpha - p + \varepsilon)\, \Gamma^{(n)}_{-aj}} \leqslant \frac{1}{(\alpha + 1 - p + \varepsilon)^{\frac{1}{p}}} \left[c_1 \| f \|^{(n-1)}_{p\, \Gamma^{(n-1)}_{-aj}} + \frac{c^2}{\varepsilon^{\frac{1}{q}}} \| f_{x_j} \|^{(n)}_{p\,(r,\, \alpha)\, \Gamma^{(n)}_{-aj}} \right],$$

which implies by virtue of (11.1) that

$$\|f\|^{(n)}_{\substack{p(r,\,\alpha-p+\varepsilon)\\ \Gamma^{(n)}_a}} \leqslant \frac{1}{(\alpha+1-p+\varepsilon)^{\frac{1}{p}}}\left[c_1\|f\|^{(n-1)}_{\substack{p\\ \Gamma^{(n-1)}_{+aj}}} + c_2\|f\|^{(n-1)}_{\substack{p\\ \Gamma^{(n-1)}_{-aj}}} + \frac{c_3}{\varepsilon^{\frac{1}{q}}}\|f_{x_j}\|^{(n)}_{\substack{p(r,\,\alpha)\\ \Gamma^{(n)}_a}}\right]. \tag{11.6}$$

[11.2]. *Suppose j is fixed* $(j = m+1,\cdots,n)$ *and* $0 \leqslant \alpha < p$. *Then there exists a constant* $c > 0$ *not depending on* f *such that*

$$\|f\|^{(n)}_{\substack{p\\ \Gamma^{(n)}_{+aj}}} \leqslant a^{\frac{1}{p}}\|f\|^{(n-1)}_{\substack{p\\ \Gamma^{(n-1)}_{+aj}}} + c\,\|f_{x_j}\|^{(n)}_{\substack{p(r,\,\alpha)\\ \Gamma^{(n)}_{+aj}}}. \tag{11.7}$$

COROLLARY.

$$\|f\|^{(n)}_{\substack{p\\ \Gamma^{(n)}_a}} \leqslant a^{\frac{1}{p}}\|f\|^{(n-1)}_{\substack{p\\ \Gamma^{(n-1)}_{+aj}}} + a^{\frac{1}{p}}\|f\|^{(n-1)}_{\substack{p\\ \Gamma^{(n-1)}_{-aj}}} + c'\,\|f_{x_j}\|^{(n)}_{\substack{p(r,\,\alpha)\\ \Gamma^{(n)}_a}}. \tag{11.8}$$

PROOF. Suppose for the sake of definiteness that $j = n$. Then

$$\|f\|^{(n)}_{\substack{p\\ \Gamma^{(n)}_{+an}}} = \Big\{\int_\Gamma\!\!\cdots\!\int dx_1\ldots dx_m \int_{-a}^{a}\!\!\cdots\int_{-a}^{a} dx_{m+1}\ldots dx_{n-1}$$

$$\times\int_0^a |f(x_1,\ldots,x_n)|^p dx_n\Big\}^{\frac{1}{p}} = \Big\{\int_\Gamma\!\!\cdots\!\int\int_{-a}^{a}\!\!\cdots\int_{-a}^{a}\int_0^a |f(x_1,\ldots,x_{n-1},a)$$

$$-\int_{x_n}^{a} f_{x_n}(x_1,\ldots,x_{n-1},t)\,dt\,|^p\,dx_1\ldots dx_n\Big\}^{\frac{1}{p}}$$

$$\leqslant\Big\{\int_0^a dx_n\int_\Gamma\!\!\cdots\!\int\int_{-a}^{a}\!\!\cdots\int_{-a}^{a}|f(x_1,\ldots,x_{n-1},a)|^p\,dx_1\ldots dx_{n-1}\Big\}^{\frac{1}{p}}$$

$$+\Big\{\int_\Gamma\!\!\cdots\!\int\int_{-a}^{a}\!\!\cdots\int_{-a}^{a}\int_0^a\Big|\int_{x_n}^{a} f_{x_n}(x_1,\ldots,x_{n-1},t)\,dt\,\Big|^p\,dx_1\ldots dx_{n-1}\Big\}^{\frac{1}{p}}$$

$$\leqslant a^{\frac{1}{p}}\|f\|^{(n-1)}_{\substack{p\\ \Gamma^{(n-1)}_{+an}}}$$

$$+\Big\{\int_\Gamma\!\!\cdots\!\int\int_{-a}^{a}\!\!\cdots\int_{-a}^{a}\int_0^a\Big|\int_{x_n}^{a} r_{(t)}^{\frac{\alpha}{p}} f_{x_n}(x_1,\ldots,x_{n-1},t)\frac{dt}{t^{\frac{\alpha}{p}}}\Big|^p\,dx_1\ldots dx_n\Big\}^{\frac{1}{p}} \leqslant$$

$$\leqslant a^{\frac{1}{p}} \|f\|_{p\, \Gamma_{+an}^{(n-1)}}^{(n-1)} + \left\{ \int \cdots \int_{\Gamma} \int_{-a}^{a} \cdots \int_{-a}^{a} \int_{0}^{a} \left(\int_{x_n}^{a} r_{(t)}^{\alpha} \right. \right.$$

$$\left. \times |f_{x_n}(x_1, \ldots, x_{n-1}, t)|^p \, dt \left(\int_{x_n}^{a} \frac{dt}{t^{\frac{\alpha q}{p}}} \right)^{\frac{p}{q}} dx_1 \ldots dx_n \right\}^{\frac{1}{p}}$$

$$\leqslant a^{\frac{1}{p}} \|f\|_{p\, \Gamma_{+an}^{(n-1)}}^{(n-1)} + \left\{ \int_0^a \left(\frac{a^{1-\frac{\alpha q}{p}} - x_n^{1-\frac{\alpha q}{p}}}{1 - \frac{\alpha q}{p}} \right)^{\frac{p}{q}} dx_n \int \cdots \int_{\Gamma} \int_{-a}^{a} \cdots \int_{-a}^{a} \int_{0}^{a} r_{(t)}^{\alpha} \right.$$

$$\left. \times |f_{x_n}(x_1, \ldots, x_{n-1}, t)|^p \, dx_1 \ldots dx_{n-1} \, dt \right\}^{\frac{1}{p}} \leqslant a^{\frac{1}{p}} \|f\|_{p\, \Gamma_{+an}^{(n-1)}}^{(n-1)}$$

$$+ \left\{ \int_0^a \left(\frac{a^{1-\frac{\alpha}{p-1}} - x_n^{1-\frac{\alpha}{p-1}}}{1 - \frac{\alpha}{p-1}} \right)^{p-1} dx_n \right\}^{\frac{1}{p}} \|f\|_{p\,(r,\,\alpha)\, \Gamma_{+an}^{(n)}}^{(n)} . \qquad (11.9)$$

If $\alpha < p - 1$, we have

$$\left(\frac{a^{1-\frac{\alpha}{p-1}} - x_n^{1-\frac{\alpha}{p-1}}}{1 - \frac{\alpha}{p-1}} \right)^{p-1} = O(1);$$

while if $p > \alpha > p - 1$, we have

$$\left(\frac{a^{1-\frac{\alpha}{p-1}} - x_n^{1-\frac{\alpha}{p-1}}}{1 - \frac{\alpha}{p-1}} \right)^{p-1} = \left[O\left(x_n^{1-\frac{\alpha}{p-1}}\right) \right]^{p-1} = O\left(\frac{1}{x_n^{1-(p-\alpha)}} \right)$$

for $x_n \to 0$. Hence in both cases there exists a constant $c > 0$ such that

$$\left\{ \int_0^a \left(\frac{a^{1-\frac{\alpha}{p-1}} - x_n^{1-\frac{\alpha}{p-1}}}{1 - \frac{\alpha}{p-1}} \right)^{p-1} dx_n \right\}^{\frac{1}{p}} < c.$$

Substituting this estimate into (11.9), we obtain inequality (11.7).

In the case $\alpha = p - 1$, by arguing in an analogous manner, we get

$$\|f\|^{(n)}_{p\,\Gamma^{(n)}_{+an}} \leqslant a^{\frac{1}{p}} \|f\|^{(n-1)}_{p\,\Gamma^{(n-1)}_{+an}} + \left\{\int_0^a \left(\ln \frac{a}{x_n}\right)^{p-1} dx_n\right\}^{\frac{1}{p}} \|f\|^{(n)}_{p(r,\alpha)\,\Gamma^{(n)}_{+an}}$$

and since $\int_0^a \ln^{p-1}(a/x_n)dx_n$ is a finite quantity, we again obtain (11.7).

[11.3]. *Suppose j is fixed $(j = m+1, \cdots, n)$, $0 \leqslant \alpha < (n-m)(p-1)$ and the m dimensional Lebesgue measure of the domain Γ is finite. Then there exists a constant $b > 0$ not depending on f such that*

$$\|f_{x_j}\|^{(n)}_{1\,\Gamma^{(n)}_{+aj}} \leqslant b\|f_{x_j}\|^{(n)}_{p(r,\alpha)\,\Gamma^{(n)}_{+aj}}.$$

COROLLARY. *Suppose $j = n$, for example, and*

$$\|f\|^{(n)}_{p(r,\alpha)\,\Gamma^{(n)}_{+an}} < \infty. \tag{11.10}$$

Then the following limit exists and is finite for almost all $(x_1, \cdots, x_{n-1})$:

$$\lim_{x_n \to 0} f(x_1, \ldots, x_{n-1}, x_n) = f(x_1, \ldots, x_{n-1}, 0) = \int_0^a f_{x_n}(x_1, \ldots, x_{n-1}, t)\,dt - f(x_1, \ldots, x_{n-1}, a). \tag{11.11}$$

PROOF. For $j = n$ we have

$$\|f_{x_n}\|^{(n)}_{1\,\Gamma^{(n)}_{+an}} = \int_\Gamma \cdots \int dx_1 \ldots dx_m \int_{-a}^{a} \cdots \int_{-a}^{a}\int_0^{a} r^{\frac{\alpha}{p}} |f_{x_n}(x_1, \ldots, x_n)| \frac{dx_{m+1} \ldots dx_n}{r^{\frac{\alpha}{p}}}$$

$$\leqslant \left\{\int_\Gamma \cdots \int \int_{-a}^{a} \cdots \int_{-a}^{a}\int_0^{a} r^\alpha |f_{x_n}(x_1, \ldots, x_n)|^p dx_1 \ldots dx_n\right\}^{\frac{1}{p}}$$

$$\times \left\{\int_\Gamma \cdots \int dx_1 \ldots dx_m \int_{-a}^{a} \cdots \int_{-a}^{a}\int_0^{a} \frac{dx_{m+1} \ldots dx_n}{r^{\frac{\alpha q}{p}}}\right\}^{\frac{1}{q}}$$

But from the fact that $\alpha q/p = \alpha/p - 1 < n - m$ and $\operatorname{mes}\Gamma < \infty$ it follows that

$$b=\left\{\int_{\Gamma}\cdots\int dx_1\dots dx_m\int_{-a}^{a}\cdots\int_{-a}^{a}\int_{0}^{a}\frac{dx_{m+1}\dots dx_n}{r^{\frac{\alpha q}{p}}}\right\}^{\frac{1}{q}}<\infty.$$

The theorem is proved for $j=n$. The cases $j=m+1,\cdots,n-1$ are considered completely analogously.

To prove the corollary we note that (11.10) for $j=n$ implies according to Fubini's theorem that the integral $\int_0^a f_{x_n}(x_1,\cdots,x_{n-1},t)dt$ is finite for almost all $(x_1,\cdots,x_{n-1})$. On the other hand, the definition of a weak derivative implies (see, for example, [4], §1.1) that for almost all $(x_1,\cdots,x_{n-1})$

$$f(x_1,\dots,x_n)=\int_{x_n}^{a} f_{x_n}(x_1,\dots,x_{n-1},t)\,dt-f(x_1,\dots,x_{n-1},a).$$

Letting x_n tend to zero when $(x_1,\cdots,x_{n-1})$ belongs to both of the indicated sets of total measure, we obtain (11.11).

Similar assertions hold, of course, for the domain $\Gamma_{-aj}^{(n)}$.

The fact that this theorem is proved for the case of a domain with a finite measure does not exclude the possibility of Γ coinciding with E^m in the sequel, since we will use only the Corollary, which has a local character.

[11.4]. *Suppose j is fixed $(j=m+1,\cdots,n)$, $0\leqslant\alpha<p$, $\alpha\neq p-1$ and $0<|h|<a-\xi$. Then there exists a constant $d_1>0$, not depending on f or ξ, such that*

$$\|\Delta_{x_j}^{(1)}(f,h)\|_{p,\Gamma_{\xi}^{(n)}}^{(n)}\leqslant d_1\left[\|f\|_{p,\Gamma_{+aj}^{(n-1)}}^{(n-1)}+\|f\|_{p,\Gamma_{-aj}^{(n-1)}}^{(n-1)}+\|f_{x_j}\|_{p(r,\alpha),\Gamma_{a}^{(n)}}^{(n)}\right]|h|^{1-\frac{\alpha}{p}}. \quad (11.12)$$

PROOF. Suppose $j=n$, for example. Then

$$\|\Delta_{x_n}^{(1)}(f,h)\|_{p,\Gamma_{\xi}^{(n)}}^{(n)}\leqslant\|\Delta_{x_n}^{(1)}(f,h)\|_{p,\Gamma_{+\xi n}^{(n)}}^{(n)}+\|\Delta_{x_n}^{(1)}(f,h)\|_{p,\Gamma_{-\xi n}^{(n)}}^{(n)}. \quad (11.13)$$

We estimate the first term in the right side; the second term is estimated completely analogously. Suppose first $\Delta x_n=x>0$. Then

$$\| \Delta_{x_n}^{(1)}(f, h) \|_{p}^{(n)}_{\Gamma_{+\xi n}^{(n)}}$$

$$= \left\{ \int_{\Gamma} \ldots \int_{-\xi}^{\xi} \ldots \int_{-\xi}^{\xi} \int_{0}^{\xi} |f(x_1, \ldots, x_{n-1}, x_n + h) - f(x_1, \ldots, x_{n-1}, x_n)|^p dx_1 \ldots dx_n \right\}^{\frac{1}{p}}$$

$$= \left\{ \int_{\Gamma} \ldots \int_{-\xi}^{\xi} \ldots \int_{-\xi}^{\xi} \int_{0}^{\xi} \left| \int_{0}^{h} f_{x_n}(x_1, \ldots, x_{n-1}, x_n + u)\, du \right|^p dx_1 \ldots dx_n \right\}^{\frac{1}{p}} \text{ (2)}$$

$$\leqslant \int_0^h \left\{ \int_{\Gamma} \ldots \int_{-\xi}^{\xi} \ldots \int_{-\xi}^{\xi} \int_{0}^{\xi} | f_{x_n}(x_1, \ldots, x_{n-1}, x_n + u)|^p dx_1 \ldots dx_n \right\}^{\frac{1}{p}} du$$

$$= \int_0^h \left\{ \int_{\Gamma} \ldots \int_{-\xi}^{\xi} \ldots \int_{-\xi}^{\xi} \int_{u}^{\xi + u} | f_{x_n}(x_1, \ldots, x_{n-1}, t)|^p dx_1 \ldots dx_{n-1}\, dt \right\}^{\frac{1}{p}} du$$

$$\leqslant \int_0^h \left\{ \int_{\Gamma} \ldots \int_{-\xi}^{\xi} \ldots \int_{-\xi}^{\xi} \int_{u}^{\xi + u} \frac{r_{(t)}^{\alpha}}{u^{\alpha}} | f_{x_n}(x_1, \ldots, x_{n-1}, t)|^p dx_1 \ldots dx_{n-1}, dt \right\}^{\frac{1}{p}} du$$

$$\leqslant \int_0^h \frac{1}{u^{\frac{\alpha}{p}}} \left\{ \int_{\Gamma} \ldots \int_{-\xi}^{\xi} \ldots \int_{-\xi}^{\xi} \int_{0}^{a} r_{(t)}^{\alpha} | f_{x_n}(x_1, \ldots, x_{n-1}, t)|^p dx_1 \ldots dx_{n-1} dt \right\}^{\frac{1}{p}} du$$

$$\leqslant \frac{p}{p - \alpha} \| f_{x_n} \|_{p(r, \alpha)}^{(n)}{}_{\Gamma_{+an}^{(n)}} h^{1 - \frac{\alpha}{p}}. \tag{11.14}$$

Suppose now $\Delta x_n = -h, h > 0$. Then

$$\| \Delta_{x_n}^{(1)}(f, -h) \|_{p}^{(n)}{}_{\Gamma_{+\xi n}^{(n)}} = \left\{ \int_{\Gamma} \ldots \int_{-\xi}^{\xi} \ldots \int_{-\xi}^{\xi} \int_{0}^{\xi} | f(x_1, \ldots, x_{n-1}, x_n - h) \right.$$

$$\left. - f(x_1, \ldots, x_{n-1}, x_n)|^p dx_1 \ldots dx_n \right\}^{\frac{1}{p}}$$

$$\leqslant \left\{ \int_{\Gamma} \ldots \int_{-\xi}^{\xi} \ldots \int_{-\xi}^{\xi} \int_{h}^{\xi} | f(x_1, \ldots, x_{n-1}, x_n - h) - f(x_1, \ldots, x_n)|^p dx_1 \ldots dx_n \right\}^{\frac{1}{p}}$$

$$+ \left\{ \int_{\Gamma} \ldots \int_{-\xi}^{\xi} \ldots \int_{-\xi}^{\xi} \int_{0}^{h} | f(x_1, \ldots, x_{n-1}, x_n - h) - f(x_1, \ldots, x_n)|^p dx_1 \ldots dx_n \right\}^{\frac{1}{p}}$$

$$= J_1 + J_2. \tag{11.15}$$

(2)We apply the generalized Minkowski inequality.

Putting $v = x_n - h$ in the first integral J_1, we get completely analogously to the proof of (11.14) that

$$J_1 = \left\{ \int_\Gamma \dots \int \int_{-\xi}^{\xi} \dots \int_{-\xi}^{\xi} \int_0^{\xi - h} | f(x_1, \dots, x_{n-1}, v+h) \right.$$

$$\left. f(x_1, \dots, x_{n-1}, v) |^p \, dx_1 \dots dx_{n-1} \, dv \right\}^{\frac{1}{p}} \leqslant \frac{p}{p-\alpha} \| f_{x_n} \|^{(n)}_{p(r,\alpha)\, \Gamma^{(n)}_{+an}} h^{1-\frac{\alpha}{p}}. \qquad (11.16)$$

The situation in regard to the second integral is more complicated. Consider first the case $p - 1 < \alpha < p$, i.e. when $0 < 1 - \alpha/p < 1/p$. We have $J_2 \leqslant J_2' + J_2''$, where

$$J_2' = \left\{ \int_\Gamma \dots \int \int_{-\xi}^{\xi} \dots \int_{-\xi}^{\xi} \int_0^{h} | f(x_1, \dots, x_{n-1}, x_n - h) |^p \, dx_1 \dots dx_n \right\}^{\frac{1}{p}}$$

$$= \left\{ \int_\Gamma \dots \int \int_{-\xi}^{\xi} \dots \int_{-\xi}^{\xi} \int_{-h}^{0} | f(x_1, \dots, x_{n-1}, v) |^p \, dx_1 \dots dx_{n-1} \, dv \right\}^{\frac{1}{p}},$$

$$J_2'' = \left\{ \int_\Gamma \dots \int \int_{-\xi}^{\xi} \dots \int_{-\xi}^{\xi} \int_0^{h} | f(x_1, \dots, x_n) |^p \, dx_1 \dots dx_n \right\}^{\frac{1}{p}}$$

Both of these integrals are estimated in the same way. We estimate the second of them, for example (when $h < 1$):

$$J_2'' = \left\{ \int_\Gamma \dots \int \int_{-\xi}^{\xi} \dots \int_{-\xi}^{\xi} \int_0^{h} | f(x_1, \dots x_{n-1}, a) \right.$$

$$\left. - \int_{x_n}^{a} f_{x_n}(x_1, \dots, x_{n-1}, t) \, dt |^p dx_1 \dots dx_n \right\}^{\frac{1}{p}}$$

$$\leqslant \left\{ \int_\Gamma \dots \int \int_{-\xi}^{\xi} \dots \int_{-\xi}^{\xi} \int_0^{h} | f(x_1, \dots, x_{n-1}, a) |^p \, dx_1 \dots dx_n \right\}^{\frac{1}{p}}$$

$$+ \left\{ \int_\Gamma \dots \int \int_{-\xi}^{\xi} \dots \int_{-\xi}^{\xi} \int_0^{h} \left| \int_{x_n}^{a} f_{x_n}(x_1, \dots, x_{n-1}, t) \, dt \right|^p dx_1 \dots dx_n \right\}^{\frac{1}{p}} \leqslant \| f \|^{(n-1)}_{p\, \Gamma^{(n-1)}_{+an}} h^{\frac{1}{p}}$$

$$+ \left\{ \int_\Gamma \dots \int \int_{-\xi}^{\xi} \dots \int_{-\xi}^{\xi} \int_0^{h} \int_{x_n}^{a} r_{(t)}^{\frac{\alpha}{p}} | f_{x_n}(x_1, \dots, x_{n-1}, t) \frac{dt}{t^{\frac{\alpha}{p}}} |^p \, dx_1 \dots dx_n \right\}^{\frac{1}{p}}$$

$$\leqslant \| f \|^{(n-1)}_{p\, \Gamma^{(n-1)}_{+an}} h^{1-\frac{\alpha}{p}} + \left\{ \int_\Gamma \dots \int \int_{-\xi}^{\xi} \dots \int_{-\xi}^{\xi} \int_0^{h} \int_{x_n}^{a} r_{(t)}^{\alpha} | f_{x_n}(x_1, \dots, x_{n-1}, t) |^p \, dt \times \right.$$

$$\times \left(\int_{x_n}^{a} \frac{dt}{t^{\frac{\alpha q}{p}}}\right)^{\frac{p}{q}} dx_1 \ldots dx_n\Bigg\}^{\frac{1}{p}} \leqslant \| f \|_{p, \Gamma^{(n-1)}_{+an}}^{(n-1)} h^{1-\frac{\alpha}{p}}$$

$$+ \Bigg\{\int_{\Gamma} \ldots \int \int_{-\xi}^{\xi} \ldots \int_{-\xi}^{\xi} \int_0^h \int_0^a r_{(t)}^{\alpha} | f_{x_n}(x_1, \ldots, x_{n-1}, t) |^p \, dt$$

$$\times \left(\frac{a^{1-\frac{\alpha q}{p}} - x_n^{1-\frac{\alpha q}{p}}}{1-\frac{\alpha q}{p}}\right)^{\frac{p}{q}} dx_1 \ldots dx_n\Bigg\}^{\frac{1}{p}} \leqslant \| f \|_{p, \Gamma^{(n-1)}_{+an}}^{(n-1)} h^{1-\frac{\alpha}{p}} \tag{11.17}$$

$$+ \Bigg\{\int_0^h \left(\frac{a^{1-\frac{\alpha q}{p}} - x_n^{1-\frac{\alpha q}{p}}}{1-\frac{\alpha q}{p}}\right)^{\frac{p}{q}} dx_n\Bigg\}^{\frac{1}{p}} \| f_{x_n} \|_{p(r, \alpha), \Gamma^{(n)}_{+an}}^{(n)},$$

where

$$\Bigg\{\int_0^h \left[\left(\frac{a^{1-\frac{\alpha q}{p}} - x_n^{1-\frac{\alpha q}{p}}}{1-\frac{\alpha q}{p}}\right)^{\frac{1}{q}}\right]^p dx_n\Bigg\}^{\frac{1}{p}} \leqslant \left(\frac{p-1}{p-1-\alpha}\right)^{\frac{p}{q}}$$

$$\times \Bigg\{\int_0^h \left(a^{\frac{1}{q}-\frac{\alpha}{p}} + x_n^{\frac{1}{q}-\frac{\alpha}{p}}\right)^p dx_n\Bigg\}^{\frac{1}{p}} (^3) \leqslant \left(\frac{p-1}{p-1-\alpha}\right)^{\frac{p}{q}}$$

$$\times \Bigg\{\left[\int_0^h a^{\frac{p}{q}-\alpha} \, dx_n\right]^{\frac{1}{p}} + \left[\int_0^h x^{\frac{p}{q}-\alpha} \, dx_n\right]^{\frac{1}{p}}\Bigg\} \leqslant \left(\frac{p-1}{p-1-\alpha}\right)^{\frac{p}{q}}$$

$$\times \left[a^{p-1-\alpha} h^{\frac{1}{p}} + \frac{1}{(p-\alpha)^{\frac{1}{p}}} h^{1-\frac{\alpha}{p}}\right] \leqslant b' h^{1-\frac{\alpha}{p}},$$

the constant b' depending only on a, p and α. Substituting this estimate into (11.17), we get

$$J_2'' \leqslant (\| f \|_{p, \Gamma^{(n-1)}_{+an}}^{(n-1)} + b' \| f_{x_n} \|_{p(r, \alpha), \Gamma^{(n)}_{+an}}^{(n)}) \, h^{1-\frac{\alpha}{p}}. \tag{11.18}$$

It remains to estimate J_2 when $0 < \alpha < p - 1$. Here we have $J_2 \leqslant J_2^* + J_2^{**}$, where

---

(3) We have made use of the inequality $(a + b)^{1/q} \leqslant a^{1/q} + b^{1/q}$, $a > 0$, $b > 0$ (see, for example, [38], inequality (2.12.2)).

$$J_2^* = \left\{ \int_\Gamma \cdots \int \int_{-\xi}^{\xi} \cdots \int_{-\xi}^{\xi} \int_0^h | f(x_1, \ldots, x_{n-1}, x_n - h) - f(x_1, \ldots, x_{n-1}, 0) |^p \, dx_1 \ldots dx_n \right\}^{\frac{1}{p}},$$

$$J_2^{**} = \left\{ \int_\Gamma \cdots \int \int_{-\xi}^{\xi} \cdots \int_{-\xi}^{\xi} \int_0^h | f(x_1, \ldots, x_{n-1}, x_n) - f(x_1, \ldots, x_{n-1}, 0) |^p \, dx_1 \ldots dx_n \right\}^{\frac{1}{p}}.$$

Both of the integrals J_2^* and J_2^{**} are also estimated in the same way. Let us estimate the second, for example. We note only that the written expressions are meaningful by virtue of Theorem [11.3], from which it follows for almost all $x_1, \cdots, x_{n-1}$ that the derivative f_{x_n} is summable, and hence

$$f(x_1, \ldots, x_{n-1}, x_n) - f(x_1, \ldots, x_{n-1}, 0) = \int_0^{x_n} f_{x_n}(x_1, \ldots, x_{n-1}, t) \, dt.$$

Therefore

$$J_2^{**} = \left\{ \int_\Gamma \cdots \int \int_{-\xi}^{\xi} \cdots \int_{-\xi}^{\xi} \int_0^h \left| \int_0^{x_n} f_{x_n}(x_1, \ldots, x_{n-1}, t) \, dt \right|^p dx_1 \ldots dx_n \right\}$$

$$\leqslant \left\{ \int_\Gamma \cdots \int \int_{-\xi}^{\xi} \cdots \int_{-\xi}^{\xi} \int_0^h \left| \int_0^{x_n} r_{(t)}^{\frac{\alpha}{p}} f_{x_n}(x_1, \ldots, x_{n-1}, t) \frac{dt}{t^{\frac{\alpha}{p}}} \right|^p dx_1 \ldots dx_n \right\}^{\frac{1}{p}}$$

$$\leqslant \left\{ \int_\Gamma \cdots \int \int_{-\xi}^{\xi} \cdots \int_{-\xi}^{\xi} \int_0^h \int_0^{x_n} r_{(t)}^{\alpha} | f_{x_n}(x_1, \ldots, x_{n-1}, t) |^p \, dt \left(\int_0^{x_n} \frac{dt}{t^{\frac{\alpha q}{p}}} \right)^{\frac{p}{q}} dx_1 \ldots dx_n \right\}^{\frac{1}{p}} \tag{11.19}$$

$$\leqslant \left\{ \int_0^h \left(\frac{x_n^{1 - \frac{\alpha q}{p}}}{1 - \frac{\alpha q}{p}} \right)^{\frac{p}{q}} dx_n \right\}^{\frac{1}{p}} \left\{ \int_\Gamma \cdots \int \int_{-a}^{a} \cdots \int_{-a}^{a} \int_0^{a} r_{(t)}^{\alpha} | f_{x_n}(x_1, \ldots, x_{n-1}, t) |^p$$

$$\times \, dx_1 \ldots dx_{n-1} \, dt \right\}^{\frac{1}{p}} = \left(\frac{p - \alpha}{p - 1 - \alpha} \right)^{\frac{1}{q}} \left\{ \int_0^h x_n^{p-1-\alpha} dx_n \right\}^{\frac{1}{p}} \| f_{x_n} \|_{p(r,\alpha)}^{(n)} {}_{\Gamma_{+an}^{(n)}}$$

$$= \left(\frac{p - 1}{p - 1 - \alpha} \right)^{\frac{1}{q}} \frac{1}{(p - \alpha)^{\frac{1}{p}}} \| f_{x_n} \|_{p(r,\alpha)}^{(n)} {}_{\Gamma_{+an}^{(n)}} h^{1 - \frac{\alpha}{p}}.$$

The theorem follows from inequalities (11.3)–(11.19).

REMARK. The case $\alpha = p - 1$ is exceptional for Theorem [11.4]. Here the following inequality holds for any $\epsilon > 0$:

$$\| \Delta_{x_j}^{(1)}(f, h) \|_{p, \Gamma_\xi^{(n)}}^{(n)} \leqslant d_1 [\| f \|_{p, \Gamma_{+aj}^{(n-1)}}^{(n-1)} + \| f \|_{p, \Gamma_{-aj}^{(n-1)}}^{(n-1)} + \| f_{x_j} \|_{p(r,\alpha), \Gamma_a^{(n)}}^{(n)}] | h |^{1 - \frac{\alpha+\varepsilon}{p}}, \qquad (11.20)$$

where the constant d_1 does not depend on f or ξ.

The proof of this inequality is carried out by successive arguments analogous to the proof of (11.12) for the case $p - 1 < \alpha < p$. We continue to have estimates (11.14) for $\Delta x_n = h > 0$ and estimates (11.15) and (11.16) for $\Delta x_n = - h$ $(h > 0)$, while in place of (11.17) we have

$$J_2'' = \left\{ \int_\Gamma \cdots \int \int_{-\xi}^{\xi} \cdots \int_{-\xi}^{\xi} \int_0^h \left| f(x_1, \dots, x_{n-1}, a) - \int_{x_n}^{a} f_{x_n}(x_1, \dots, x_{n-1}, t)\, dt \right|^p dx_1 \dots dx_n \right\}^{\frac{1}{p}}$$

$$\leqslant \| f \|_{p, \Gamma_{+an}^{(n-1)}}^{(n-1)} h^{\frac{1}{p}} + \left\{ \int_\Gamma \cdots \int \int_{-\xi}^{\xi} \cdots \int_{-\xi}^{\xi} \int_0^h \left| \int_{x_n}^{a} r_{(t)}^{\frac{p-1}{p}} f_{x_n}(x_1, \dots, x_{n-1}, t) \frac{dt}{t^{\frac{p-1}{p}}} \right|^p dx_1 \dots dx_n \right\}^{\frac{1}{p}}$$

$$\leqslant \| f \|_{p, \Gamma_{+an}^{(n-1)}}^{(n-1)} h^{\frac{1}{p}} + \left\{ \int_\Gamma \cdots \int \int_{-\xi}^{\xi} \cdots \int_{-\xi}^{\xi} \int_0^h \int_{x_n}^{a} r_{(t)}^{p-1} \left| f_{x_n}(x_1, \dots, x_{n-1}, t) \right|^p dt \left(\int_{x_n}^{a} \frac{dt}{t} \right)^{\frac{p}{q}} \right.$$

$$\left. \times dx_1 \dots dx_n \right\}^{\frac{1}{p}} \leqslant \| f \|_{p, \Gamma_{+an}^{(n-1)}}^{(n-1)} h^{\frac{1}{p}} + \left\{ \int_0^h \left(\ln \frac{\xi}{x_n} \right)^{\frac{p}{q}} dx_n \right\}^{\frac{1}{p}} \| f_{x_n} \|_{p(r, p-1), \Gamma_{+an}^{(n)}}^{(n)}.$$

For any $\epsilon \in (0, 1)$

$$\left\{ \int_0^h \left(\ln \frac{\xi}{x_n} \right)^{\frac{p}{q}} dx_n \right\}^{\frac{1}{p}} \leqslant b'' \left\{ \int_0^h \frac{dx_n}{x_n^{\varepsilon}} \right\}^{\frac{1}{p}} = \frac{b''}{(1-\varepsilon)^{\frac{1}{p}}} h^{\frac{1-\varepsilon}{p}} = \frac{b''}{(1-\varepsilon)^{\frac{1}{p}}} | h |^{1 - \frac{\alpha+\varepsilon}{p}}, \quad \alpha = p - 1,$$

i.e.

$$J_2'' \leqslant \| f \|_{p, \Gamma_{+an}^{(n-1)}}^{(n-1)} h^{1 - \frac{\alpha}{p}} + \frac{b''}{(1-\varepsilon)^{\frac{1}{p}}} \| f_{x_n} \|_{p(r, \alpha), \Gamma_{+an}^{(n)}}^{(n)} h^{1 - \frac{\alpha+\varepsilon}{p}}, \quad \alpha = p - 1,$$

where the constant b'' does not depend on f, ξ or ϵ. Combining this estimate with (11.14), (11.13) and (11.16), we obtain (11.20).

We note that (11.12) and (11.20) imply that inequality (11.20) holds under the assumptions of Theorem [11.4] for any $\alpha \in (0, p)$.

Before proceeding to an estimate of the norm of a difference with respect to the variables $x_1, \cdots, x_m$, we prove a preliminary lemma.

[11.5.1]. *Suppose* i *and* j *are fixed* $(i = 1, \cdots, m; j = m+1, \cdots, n)$, $0 < h < \eta$ *and* $0 < \alpha < p$. *Then*

$$I(h) = \left\{ \int_{\Gamma_\eta} \cdots \int dx_1 \ldots dx_m \int_{-\xi}^{\xi} \cdots \int_{-\xi}^{\xi} dx_{m+1} \ldots dx_{j-1} dx_{j+1} \ldots dx_n \int_h^{\xi} |\Delta_{x_i}^{(1)}(f, h)|^p dx_j \right\}^{\frac{1}{p}}$$

$$\leqslant \| f_{x_i} \|^{(n)}_{\substack{p(r,\alpha) \\ \Gamma^{(n)}_{+aj}}} h^{1-\frac{\alpha}{p}}.$$

PROOF. Suppose for the sake of definiteness that $i = 1$ and $j = n$. Then

$$I(h) = \left\{ \int_{\Gamma_\eta} \cdots \int \int_{-\xi}^{\xi} \cdots \int_{-\xi}^{\xi} \int_h^{\xi} \left| \int_0^h f_{x_1}(x_1 + u, x_2, \ldots, x_n)\, du \right|^p dx_1 \ldots dx_n \right\}^{\frac{1}{p}}$$

$$\leqslant \int_0^h \left\{ \int_{\Gamma_\eta} \cdots \int \int_{-a}^{a} \cdots \int_{-a}^{a} \int_h^{a} |f_{x_1}(x_1 + u, x_2, \ldots, x_n)|^p dx_1 \ldots dx_n \right\}^{\frac{1}{p}} du$$

$$\leqslant \int_0^h \left\{ \int_{\Gamma} \cdots \int \int_{-a}^{a} \cdots \int_{-a}^{a} \int_h^{a} |f_{x_1}(t, x_2, \ldots x_n)|^p dt dx_2 \ldots dx_n \right\}^{\frac{1}{p}} du$$

$$\leqslant h \left\{ \int_{\Gamma} \cdots \int \int_{-a}^{a} \int_h^{a} \frac{r^\alpha}{h^\alpha} |f_{x_1}(x_1, x_2, \ldots, x_n)|^p dx_1 \ldots dx_n \right\}^{\frac{1}{p}}$$

$$\leqslant \left\{ \int_{\Gamma} \cdots \int \int_{-a}^{a} \cdots \int_{-a}^{a} \int_0^{a} r^\alpha |f_{x_1}(x_1, x_2, \ldots, x_n)|^p dx_1 \ldots dx_n \right\}^{\frac{1}{p}} h^{1-\frac{\alpha}{p}}$$

$$= \| f_{x_1} \|^{(n)}_{\substack{p(r,\alpha) \\ \Gamma^{(n)}_{+an}}} h^{1-\frac{\alpha}{p}}.$$

[11.5]. *Suppose* i *and* j *are fixed* $(i = 1, \cdots, m; j = m+1, \cdots, n)$, $0 < |h| < \eta$ *and* $0 < \alpha < p$. *Then for any* $\epsilon > 0$ *there exists a constant* $d_2 > 0$ *not depending on* f *or* η *such that*

$$\|\Delta^{(1)}_{x_i}(f,h)\|^{(n)}_{p,\Gamma^{(n)}_{\eta\xi}} \leqslant d_2\big[\|f\|^{(n-1)}_{p,\Gamma^{(n-1)}_{+aj}} + \|f\|^{(n-1)}_{p,\Gamma^{(n-1)}_{-aj}} + \|f_{x_i}\|^{(n)}_{p(r,\alpha),\Gamma^{(n)}_a} + \|f_{x_j}\|^{(n)}_{p(r,\alpha),\Gamma^{(n)}_a}\big]|h|^{1-\frac{\alpha+\varepsilon}{p}}, \tag{11.21}$$

where $\xi = a - \eta > 0$, *it being possible to take* $\epsilon = 0$ *in the right side when* $\alpha \neq p-1$.

PROOF. Suppose for the sake of definiteness that $j = n$ and $i = 1$. It can be assumed without loss of generality that $0 < h < \xi$, since when $\Delta x_n = -h$, for example, it suffices to make the change of variable $v = x_1 - h$, after which all of the arguments carried out below remain in force. Thus suppose $0 < h < \xi$. We have

$$\|\Delta^{(1)}_{x_1}(f,h)\|^{(n)}_{p,\Gamma^{(n)}_{\eta\xi}} \leqslant \|\Delta^{(1)}_{x_1}(f,h)\|^{(n)}_{p,\Gamma^{(n)}_{\eta(+\xi)n}} + \|\Delta^{(1)}_{x_1}(f,h)\|^{(n)}_{p,\Gamma^{(n)}_{\eta(-\xi)n}}.$$

Both of the terms in the right side are estimated in the same way. We estimate the first of them:

$$\|\Delta^{(1)}_{x_1}(f,h)\|^{(n)}_{p,\Gamma^{(n)}_{\eta(+\xi)n}} \leqslant I(h) + \tilde{I}(h),$$

where the integral

$$I(h) = \left\{\int_{\Gamma_\eta}\cdots\int\int_{-\xi}^{\xi}\cdots\int_{-\xi}^{\xi}\int_h^{\xi}|\Delta^{(1)}_{x_1}(f,h)|^p\,dx_1\ldots dx_n\right\}^{\frac{1}{p}}$$

has been estimated in Lemma [11.5.1]:

$$I(h) \leqslant \|f_{x_1}\|^{(n)}_{p(r,\alpha),\Gamma^{(n)}_{+an}}|h|^{1-\frac{\alpha}{p}},$$

while

$$\tilde{I}(h) = \left\{\int_{\Gamma_\eta}\cdots\int\int_{-\xi}^{\xi}\cdots\int_{-\xi}^{\xi}\int_0^{h}|\Delta^{(1)}_{x_1}(f,h)|^p\,dx_1\ldots dx_n\right\}^{\frac{1}{p}} \leqslant I_1 + I_2 + I_3,$$

where

$$I_1=\left\{\int_{\Gamma_\eta}\cdots\int\int_{-\xi}^{\xi}\cdots\int_{-\xi}^{\xi}\int_0^h|f(x_1+h,\ x_2,\ldots,\ x_n+h)\right.$$

$$\left.-f(x_1+h,x_2,\ldots,x_{n-1},x_n)|^p\,dx_1,\ \ldots\ dx_n\right\}^{\frac{1}{p}}$$

$$\leqslant\left\{\int_{\Gamma}\cdots\int\int_{-\xi}^{\xi}\cdots\int_{-\xi}^{\xi}\int_0^{\xi}|f(x_1,\ x_2,\ \ldots,\ x_{n-1},\ x_n+h)\right.$$

$$\left.-f(x_1,\ x_2,\ \ldots,\ x_{n-1},\ x_n)|^p\,dx_1\ldots dx_n\right\}^{\frac{1}{p}}$$

$$\leqslant d'\,[\,\|f\|_{\substack{p\\ \Gamma_{+an}^{(n-1)}}}^{(n-1)}+\|f\|_{\substack{p\\ \Gamma_{-an}^{(n-1)}}}^{(n-1)}+\|f_{x_n}\|_{\substack{p(r,\alpha)\\ \Gamma_a^{(n)}}}^{(n)}]\,|h|^{1-\frac{\alpha+\varepsilon}{p}}$$

by virtue of (11.20),

$$I_2=\left\{\int_{\Gamma_\eta}\cdots\int\int_{-\xi}^{\xi}\cdots\int_{-\xi}^{\xi}\int_0^h|f(x_1+h,\ x_2,\ \ldots,\ x_{n-1},\ x_n+h)\right.$$

$$\left.-f(x_1,\ x_2,\ \ldots,\ x_{n-1},\ x_n+h)|^p\,dx_1\ldots dx_n\right\}^{\frac{1}{p}}$$

$$\leqslant\left\{\int_{\Gamma_\eta}\cdots\int\int_{-\xi}^{\xi}\cdots\int_{-\xi}^{\xi}\int_h^{2h}|f(x_1+h,\ x_2,\ \ldots,\ x_{n-1},\ t)\right.$$

$$\left.-f(x_1,\ \ldots,x_{n-1},\ t)|^p\,dx_1\ldots dx_{n-1}\,dt\right\}^{\frac{1}{p}}\leqslant\|f_{x_1}\|_{\substack{p(r,\alpha)\\ \Gamma_{+an}^{(n)}}}^{(n)}h^{1-\frac{\alpha}{p}}$$

by virtue of Lemma [11.5.1], and

$$I_3=\left\{\int_{\Gamma_\eta}\cdots\int\int_{-\xi}^{\xi}\cdots\int_{-\xi}^{\xi}\int_0^h|f(x_1,\ldots,x_{n-1},\ x_n+h)-f(x_1,\ldots,x_n)|^p dx_1\ldots dx_n\right\}^{\frac{1}{p}}$$

$$\leqslant\left\{\int_{\Gamma_\eta}\cdots\int\int_{-\xi}^{\xi}\cdots\int_{-\xi}^{\xi}\int_0^{\xi}|f(x_1,\ldots,x_{n-1},x_n+h)-f(x_1,\ldots,x_n)|^p dx_1\ldots dx_n\right\}^{\frac{1}{p}}$$

$$\leqslant d'\,[\,\|f\|_{\substack{p\\ \Gamma_{+an}^{(n-1)}}}^{(n-1)}+\|f\|_{\substack{p\\ \Gamma_{-an}^{(n-1)}}}^{(n-1)}+\|f_{x_n}\|_{\substack{p(r,\alpha)\\ \Gamma_a^{(n)}}}^{(n)}]\,|h|^{1-\frac{\alpha+\varepsilon}{p}}$$

by virtue of (11.20) again. The theorem is proved.

Let us now waive the initial assumption concerning the existence for almost all $(x_1, \cdots, x_{j-1}, 0, x_{j+1}, \quad , x_n)$ of a limit of the function f for $x_j \to a - 0$ (and for $x_j \to -a + 0$), and ascertain when it is valid.

[11.6]. *Suppose* j *is fixed* $(j = m + 1, \cdots, n)$, Γ *is a bounded domain,* $\alpha \geqslant 0$ *and*

$$\| f \|_{p}^{(n)}{}_{\Gamma^{(n)}_{+aj}} < \infty, \qquad \| f_{x_j} \|_{p(r,\alpha)}^{(n)}{}_{\Gamma^{(n)}_{+aj}} < \infty. \tag{11.22}$$

Then f *has a value on* $\Gamma^{(n-1)}_{+aj}$ *to which it converges both in the mean and almost everywhere, and*

$$\| f \|_{p}^{(n-1)}{}_{\Gamma^{(n-1)}_{+aj}} < \infty. \tag{11.23}$$

An analogous assertion holds, of course, for the domain $\Gamma^{(n)}_{-aj}$.

PROOF. Suppose for the sake of definiteness that $j = n$. By virtue of the existence of the weak derivative f_{x_n} in $\Gamma^{(n)}_{+an}$ the function f can be so modified on a set of zero measure that for almost all points $P = (x_1, \cdots, x_{n-1}) \in \Gamma$ it will be absolutely continuous in x_n on segments parallel to the Ox_n-axis, projecting into the point P and lying entirely in the domain $\Gamma^{(n)}_{+an}$. Let $0 < a' < a$ and

$$G = \Gamma^{(n)}_{+an} \cap \{(x_1, \dots x_n) : x_n > a'\}.$$

From the second condition of (11.22) and the fact that $r > a'$ on G it follows that $\|f_{x_n}\|_{p}^{(n)}{}_{G} < \infty$, which implies by virtue of the boundedness of Γ that

$\|f_{x_n}\|_{1}^{(n)}{}_{G} < \infty$, and hence that f_{x_n} is summable with respect to x_n on the segments $[a', a]$ for almost all $(x_1, \cdots, x_{n-1})$. Therefore the following limit exists and is finite for almost all $(x_1, \cdots, x_{n-1})$:

$$\lim_{x_n \to a-0} f(x_1, \dots, x_{n-1}, x_n) = f(x_1, \dots, x_{n-1}, a') + \int_{a'}^{a} f_{x_n}(x_1, \dots, x_{n-1}, t)\, dt$$
$$= f(x_1, \dots, x_{n-1}, a).$$

We show that it is also a limit in the mean:

$$\int_\Gamma \cdots \int_a \int_{-a}^{a} \cdots \int_{-a}^{a} |f(x_1, \ldots, x_{n-1}, a) - f(x_1, \ldots, x_{n-1}, x_n)|^p \, dx_1 \ldots dx_{n-1}$$

$$= \int_\Gamma \cdots \int \int_{-a}^{a} \cdots \int_{-a}^{a} \left| \int_{x_n}^{a} f_{x_n}(x_1, \ldots, x_{n-1}, t)\, dt \right|^p dx_1 \ldots dx_{n-1}$$

$$\leq (a - x_n)^{\frac{p}{q}} \int_\Gamma \cdots \int \int_{-a}^{a} \cdots \int_{-a}^{a} |f_{x_n}(x_1, \ldots, x_{n-1}, t)|^p \, dx_1 \ldots dx_{n-1}\, dt \to 0$$

for $x_n \to a$. Thus f converges in the mean to the value $f(x_1, \cdots, x_{n-1}, a)$.

Further, from the first condition of (11.22) it follows that there exists a number $\xi_0 \in (a', a)$ such that

$$\| f \|_{p, G_{\xi_0}}^{(n-1)} < \infty, \text{ where } G_{\xi_0} = G \cap \{(x_1, \ldots, x_n) : x_n = \xi_0\};$$

Hence

$$\| f \|_{p, \Gamma_{+an}^{(n-1)}}^{(n-1)} = \left\{ \int_\Gamma \cdots \int \int_{-a}^{a} \cdots \int_{-a}^{a} |f(x_1, \ldots, x_{n-1}, a)|^p \, dx_1 \ldots dx_{n-1} \right\}^{\frac{1}{p}}$$

$$= \left\{ \int_\Gamma \cdots \int \int_{-a}^{a} \cdots \int_{-a} \left| f(x_1, \ldots, x_{n-1}, \xi_0) + \int_{\xi_0}^{a} f_{x_n}(x_1, \ldots, x_{n-1}, t)\, dt \right|^p dx_1 \ldots dx_{n-1} \right\}^{\frac{1}{p}}$$

$$\leq \| f \|_{p, G_{\xi_0}}^{(n-1)} + \left\{ \int_\Gamma \cdots \int \int_{-a}^{a} \cdots \int_{-a} \left| \int_{\xi_0}^{a} f_{x_n}(x_1, \ldots, x_{n-1}, t)\, dt \right|^p dx_1 \ldots dx_{n-1} \right\}^{\frac{1}{p}}$$

$$\| f \|_{p, G_{\xi_0}}^{(n-1)} + (a - \xi_0)^{\frac{1}{q}} \left\{ \int_\Gamma \cdots \int \int_{-a}^{a} \cdots \int_{-a}^{a} \int_{\xi_0}^{a} |f_{x_n}(x_1, \ldots, x_{n-1}, t)|^p \, dx_1 \ldots dx_{n-1}\, dt \right\}^{\frac{1}{p}}$$

$$\leq \| f \|_{p, G_{\xi_0}}^{(n-1)} + (a - \xi_0)^{\frac{1}{q}} \| f_{x_n} \|_{p, G}^{(n)} < \infty.$$

REMARK. The first of the conditions (11.22), which was used in the proof of [11.6] only to obtain (11.23), can be replaced by the condition $\| f \|_{p, G}^{(n)} < \infty$. We note that this condition is a consequence of the following assumption, for example:

$$\sum_{i=1}^{n} \| f_{x_i} \|_{p, G}^{(n)} < \infty.$$

In the case $n=1$ this means that $\int_{a'}^{a}|f'(x)|^p dx<\infty$ implies $\int_{a'}^{a}|f(x)|^p dx<\infty$, which we prove by noting that

$$\left\{\int_{a'}^{a}|f(x)|^p dx\right\}^{\frac{1}{p}}=\left\{\int_{a'}^{a}\left|f(a')+\int_{a'}^{x}f'(t)\,dt\right|^p dx\right\}^{\frac{1}{p}}$$
$$\leqslant|f(a')|(a-a')^{\frac{1}{p}}+\left\{\int_{a'}^{a}\left|\int_{a'}^{x}f'(t)\,dt\right|^p dx\right\}^{\frac{1}{p}}\leqslant|f(a')|(a-a')^{\frac{1}{p}}$$
$$+\left\{\int_{a'}^{a}dx\int_{a'}^{x}|f'(t)|^p dt\right\}^{\frac{1}{p}}(a-a')^{\frac{1}{q}}\leqslant|f(a')|(a-a')^{\frac{1}{p}}+(a-a')\,\|f'\|_p^{(1)}<\infty.$$

Suppose our assertion has been proved for $n-1$ and we wish to prove it for n. From the condition $\sum_{i=1}^{n}\|f_{x_i}\|_{p,G}^{(n)}<\infty$ it follows that there exists a ξ_0 such that $\sum_{i=1}^{n-1}\|f_{x_i}\|_{p,G_{\xi_0}}^{(n-1)}<\infty$, $a'<\xi_0<a$. But this implies by assumption that $\|f\|_{p,G_{\xi_0}}^{(n-1)}<\infty$. Hence, in exactly the same way as in the case $n=1$ (in essence, analogously to the proof of Theorem [11.2] for $\alpha=0$) it follows that

$$\|f\|_{p,G}^{(n)}\leqslant\gamma_1\|f\|_{p,G_{\xi_0}}^{(n-1)}+\gamma_2\|f_{x_n}\|_{p,G}^{(n)}<\infty,$$

where the constants $\gamma_1>0$ and $\gamma_2>0$ do not depend on f.

[11.7] (IMBEDDING THEOREM FOR WEIGHTED FUNCTION CLASSES). *Suppose* $1\leqslant p<\infty$, *a function* f *is defined on the domain* $\Gamma_a^{(n)}$ *and all of its weak derivatives up to order* $s+1$ *inclusively exist on the set* $\Gamma_a^{(n)}\setminus E^m$, *the following inequalities holding for some* $\sigma\in[sp,(s+1)p)$:

$$\|f^{(s+1)}_{x_1^{s_1}\ldots x_n^{s_n}}\|_{p(r,\sigma),\Gamma_a^{(n)}}^{(n)}<\infty\quad \textit{for all}\quad s_1+\ldots+s_n=s+1. \tag{11.24}$$

Suppose, further, $\sigma=sp+\alpha$, s *is a nonnegative integer and* $0\leqslant\alpha<p$. *Then for any* $\epsilon\in(0,(1+s^{-1})(p-\alpha))$

$$\|f^{(s-t)}_{x_1^{t_1}\ldots x_n^{t_n}}\|_{p\left[r,\sigma-(t+1)p+\frac{t+1}{s+1}\epsilon\right],\Gamma_a^{(n)}}^{(n)}\leqslant M_t,\quad t=0,1,\ldots,s-1,s;\ t_1+\ldots+t_n=s-t, \tag{11.25}$$

$$\| f \|_{\substack{p \\ \Gamma_a^{(n)}}}^{(n)} \leqslant M_s, \tag{11.26}$$

where

$$\begin{aligned} M_t \leqslant A_t \sum_{l=0}^{t} \sum_{l_1+\ldots+l_n=s-l} \Big[\| f^{(s-l)}_{x_1^{l_1}\ldots x_n^{l_n}} \|^{(n-1)}_{\substack{p \\ \Gamma^{(n-1)}_{-an}}} + \| f^{(s-l)}_{x_1^{l_1}\ldots x_n^{l_n}} \|^{(n-1)}_{\substack{p \\ \Gamma^{(n-1)}_{+an}}} \Big] \\ + B_t \sum_{s_1+\ldots+s_n=s+1} \| f^{(s+1)}_{x_1^{s_1}\ldots x_n^{s_n}} \|_{\substack{p(r,\sigma) \\ \Gamma_a^{(n)}}}, \\ t = 0,\ 1,\ \ldots,\ s; \end{aligned} \tag{11.27}$$

the constants A_t and B_t not depending on f. Finally,

$$f \in H_{p(n)}^{1-\frac{\alpha+\varepsilon}{p}} (M_s,\ \Gamma_a^{(n)}) \tag{11.28}$$

(ϵ *can be set equal to zero in the case* $s = 0$ *and* $\alpha \neq p-1$).

Proof. Suppose the numbers $t_1, \cdots, t_n$ are fixed. From (11.24), as a special case, we have

$$\| f^{(s+1)}_{x_1^{t_1}\ldots x_n^{t_n+t+1}} \|^{(n)}_{p(r,\sigma)} < \infty, \quad t_1 + \ldots + t_n = s - t.$$

Applying inequality (11.6) with $j = n$ and ϵ replaced by $\epsilon/(s+1)$ to the derivative $f^{(s+1)}_{x_1^{t_1}, \cdots x_n^{t_n+t+1}}$, we get

$$\begin{aligned} &\| f^{(s)}_{x_1^{t_1}\ldots x_n^{t_n+t}} \|^{(n)}_{\substack{p\left(r,\ \sigma - p + \frac{\varepsilon}{s+1}\right) \\ \Gamma_a^{(n)}}} \\ \leqslant c_1' \Big[&\| f^{(s)}_{x_1^{t_1}\ldots x_n^{t_n+t}} \|^{(n-1)}_{\substack{p \\ \Gamma^{(n-1)}_{-an}}} + \| f^{(s)}_{x_1^{t_1}\ldots x_n^{t_n+t}} \|^{(n-1)}_{\substack{p \\ \Gamma^{(n-1)}_{+an}}} \Big] + c_2' \| f^{(s+1)}_{x_1^{t_1}\ldots x_n^{t_n+t+1}} \|^{(n)}_{\substack{p(r,\sigma) \\ \Gamma_a^{(n)}}}, \end{aligned}$$

where the constants c_1' and c_2' do not depend on f. Successively carrying out analogous arguments t times, we induce that

$$\| f^{(s-t)}_{x_1^{t_1}\dots x_n^{t_n}} \|^{(n)}_{p\left(r,\sigma-(t+1)p+\frac{t+1}{s+1}\varepsilon\right)} \leqslant M'_t,$$

where

$$M'_t \leqslant A'_t \sum_{l=0}^{t} \Big(\| f^{s-t+l}_{x_1^{t_1}\dots x_n^{t_n+l}} \|^{(n-1)}_{p,\,\Gamma^{(n-1)}_{-an}} + \| f^{s-t+l}_{x_1^{t_1}\dots x_n^{t_n+l}} \|^{(n-1)}_{p,\,\Gamma^{(n-1)}_{+an}} \Big)$$

$$+ B'_t \| f^{(s+1)}_{x_1^{t_1}\dots x_n^{t_n+t+1}} \|^{(n)}_{p(r,\sigma),\,\Gamma^{(n)}_a}, \quad t = 0,\ 1,\ \dots,\ s-1.$$

This proves (11.25) and (11.27), and in a somewhat sharpened form. It follows, in particular, when $t = s - 1$ that

$$\| f_{x_i} \|^{(n)}_{p\left(r,\,\alpha+\frac{s}{s+1}\varepsilon\right),\,\Gamma^{(n)}_a} \leqslant M_{s-1}, \quad i = 1,\ 2,\ \dots,\ n,$$

and from inequality (11.8) with $j = n$ that

$$\begin{aligned} \| f \|^{(n)}_{p,\,\Gamma^{(n)}_a} &\leqslant a^{\frac{1}{p}} \| f \|^{(n-1)}_{p,\,\Gamma^{(n-1)}_{-an}} + a^{\frac{1}{p}} \| f \|^{(n-1)}_{p,\,\Gamma^{(n-1)}_{+an}} + c' \| f_{x_n} \|^{(n)}_{p\left(r,\,\alpha+\frac{s}{s+1}\varepsilon\right)} \\ &\leqslant a^{\frac{1}{p}} \Big(\| f \|^{(n-1)}_{p,\,\Gamma^{(n-1)}_{-an}} + \| f \|^{(n-1)}_{p,\,\Gamma^{(n-1)}_{+an}} \Big) + c' M_{s-1}. \end{aligned} \tag{11.29}$$

Further, from inequalities (11.20) and (11.21) with ϵ and α replaced by $\epsilon/(s + 1)$ and $\alpha + s\epsilon/(s + 1)$ respectively we get

$$\begin{aligned} &\| \Delta^{(1)}_{x_i} (f;\, h) \|^{(n)}_{p,\,\Gamma^{(n)}_{\eta\xi}} \\ &\leqslant d\Big[\| f \|^{(n-1)}_{p,\,\Gamma^{(n-1)}_{-an}} + \| f \|^{(n-1)}_{p,\,\Gamma^{(n-1)}_{+an}} + \| f_{x_i} \|^{(n)}_{p\left(r,\alpha+\frac{s}{s+1}\varepsilon\right),\,\Gamma^{(n)}_a} + \| f_{x_n} \|_{p\left(r,\alpha+\frac{s}{s+1}\varepsilon\right),\,\Gamma^{(n)}_a} \Big] |h|^{1-\frac{\alpha+\varepsilon}{p}} \\ &\leqslant d\Big[\| f \|^{(n-1)}_{p,\,\Gamma^{(n-1)}_{-an}} + \| f \|^{(n-1)}_{p,\,\Gamma^{(n-1)}_{+an}} + 2M'_{s-1}\Big] |h|^{1-\frac{\alpha+\varepsilon}{p}}, \quad i = 1,\ 2,\ \dots,\ n, \end{aligned} \tag{11.30}$$

where $d = \max(d_1, d_2)$. Denoting by M_s the larger of two quantities one of which is the right side of (11.29) while the other is the coefficient of $|h|^{1-(\alpha+\epsilon)/p}$ in (11.30), we obtain (11.26) and (11.28). It should only be noted in this connection that (11.24) implies by virtue of Theorem [11.6] that all of the M_t are finite $(t = 0, 1, \cdots, s)$.

As can be seen from the cited proof, the assertion of the theorem remains valid under a certain weakening of condition (11.24); namely, it suffices that this condition be satisfied not for all derivatives of order $s+1$ but only for those needed to permit the above use of inequality (11.6) in deriving (11.25) for $t = 0$. For example, when $s \geqslant 1$ it is obviously sufficient if condition (11.24) is satisfied for all mixed derivatives.

We did not state this in the formulation of the theorem for the sake of simplicity in the presentation.

We now study the case $p = \infty$.

[11.8]. *Suppose* i *and* j *are fixed* $(i = 1, \cdots, m; j = m+1, \cdots, n)$, $p = \infty$, $0 < \epsilon < 1$ *and* $|h| < \eta$. *Then there exist constants* $c_k > 0$, $k = 1, 2, 3, 4$, *not depending on* f *or* η, *such that*

$$\| r^{\alpha-1+\epsilon} f \|^{(n)}_{\infty,\, \Gamma_a^{(n)}} \leqslant c_1 \left[\| f \|^{(n-1)}_{\infty,\, \Gamma_{-aj}^{(n-1)}} + \| f \|^{(n-1)}_{\infty,\, \Gamma_{+aj}^{(n-1)}} + \| r^{\alpha} f_{x_j} \|^{(n)}_{\infty,\, \Gamma_a^{(n)}} \right], \tag{11.31}$$

for $\alpha \geqslant 1$ *and*

$$\| f \|^{(n)}_{\infty,\, \Gamma_a^{(n)}} \leqslant c_2 \left[\| f \|^{(n-1)}_{\infty,\, \Gamma_{-aj}^{(n-1)}} + \| f \|^{(n-1)}_{\infty,\, \Gamma_{+aj}^{(n-1)}} + \| r^{\alpha} f_{x_j} \|^{(n)}_{\infty,\, \Gamma_a^{(n)}} \right], \tag{11.32}$$

$$\| \Delta^{(1)}_{x_j} (f, h) \|^{(n)}_{\infty,\, \Gamma_\xi^{(n)}} \leqslant c_3 \| r^{\alpha} f_{x_j} \|^{(n)}_{\infty,\, \Gamma_a^{(n)}} |h|^{1-\alpha}, \tag{11.33}$$

$$\| \Delta^{(1)}_{x_i} (f, h) \|^{(n)}_{\infty,\, \Gamma_{\tau,\xi}^{(n)}} < c_4 \left[\| r^{\alpha} f_{x_i} \|^{(n)}_{\infty,\, \Gamma_a^{(n)}} + \| r^{\alpha} f_{x_j} \|^{(n)}_{\infty,\, \Gamma_a^{(n)}} \right] |h|^{1-\alpha} \tag{11.34}$$

for $0 < \alpha < 1$.

We will assume that $i=1, j=n$ and $(x_1, \cdots, x_n) \in \Gamma_a^{(n)}$.

PROOF OF (11.31). Noting that $r < ma$ while $r < r_{(t)}$ for $0 < x_n \leqslant t \leqslant a$ (for the notation see the proof of [11.1]), we have for $x_n > 0$

$$r^{\alpha-1+\varepsilon} f(x_1, \ldots, x_n)| = \left| r^{\alpha-1+\varepsilon} \left[f(x_1, \ldots, x_{n-1}, a) - \int_{x_n}^{a} f_{x_n}(x_1, \ldots, x_{n-1}, t), dt \right] \right|$$

$$\leqslant | r^{\alpha-1+\varepsilon} f(x_1, \ldots, x_{n-1}, a) | + \int_{x_n}^{a} \frac{| r_{(t)}^{\alpha} f_{x_n}(x_1, \ldots, x_{n-1}, t) |}{r_{(t)}^{1-\varepsilon}} dt$$

$$\leqslant (ma)^{\alpha-1+\varepsilon} \| f \|_{\Gamma_{+an}^{(n-1)}}^{(n-1)}{}_{\infty} + \| r^{\alpha} f_{x_n} \|_{\Gamma_{+an}^{(n)}}^{(n)}{}_{\infty} \int_0^a \frac{dt}{t^{1-\varepsilon}}$$

$$\leqslant (ma)^{\alpha} \| f \|_{\Gamma_{+an}^{(n-1)}}^{(n-1)}{}_{\infty} + \frac{a}{\varepsilon} \| r^{\alpha} f_{x_n} \|_{\Gamma_{+an}^{(n)}}^{(n)}{}_{\infty};$$

Taking the supremum of the left side, we obtain (11.31) for $x_n > 0$. The case $x_n < 0$ is analyzed analogously.

PROOF OF (11.32). For $x_n > 0$

$$| f(x_1, \ldots, x_n) | = \left| f(x_1, \ldots, x_{n-1}, a) - \int_{x_n}^{a} f_{x_n}(x_1, \ldots, x_{n-1}, t)\, dt \right|$$

$$\leqslant | f(x_1, \ldots, x_{n-1}, a) | + \int_{x_n}^{a} r_{(t)}^{\alpha} | f_{x_n}(x_1, \ldots, x_{n-1}, t) | \frac{dt}{t^{\alpha}}$$

$$\leqslant \| f \|_{\Gamma_{+an}^{(n-1)}}^{(n-1)}{}_{\infty} + \| r^{\alpha} f_{x_n} \|_{\Gamma_{a}^{(n)}}^{(n)}{}_{\infty} \int_0^a \frac{dt}{t^{\alpha}}$$

$$= \| f \|_{\Gamma_{+an}^{(n-1)}}^{(n-1)}{}_{\infty} + \frac{a^{1-\alpha}}{1-\alpha} \| r^{\alpha} f_{x_n} \|_{\Gamma_{+an}^{(n)}}^{(n)}{}_{\infty},$$

The case $x_n < 0$ is analyzed analogously.

PROOF OF (11.33). Suppose first $x_n > 0$ and $h > 0$. Then

$$|\Delta_{x_n}^{(1)}(f, h)| = \left| \int_{x_n}^{x_n+h} f_{x_n}(x_1, \ldots, x_{n-1}, t)\, dt \right| \leqslant \int_{x_n}^{x_n+h} r_{(t)}^{\alpha} |f_{x_n}(x_1, \ldots, x_{n-1}, t)| \frac{dt}{t^{\alpha}}$$

$$\leqslant \frac{1}{1-\alpha} \| r^{\alpha} f_{x_n} \|_{\infty}^{(n)} {}_{1_{+a}^{(n)}} [(x_n + h)^{1-\alpha} - x_n^{1-\alpha}] \leqslant \frac{2^{1-\alpha}}{1-\alpha} \| r^{\alpha} f_{x_n} \|_{\infty}^{(n)} {}_{\Gamma_a^{(n)}} h^{1-\alpha}$$

For consider the function $\varphi(x) = (x + h)^{1-\alpha} - x^{1-\alpha}, 0 < \alpha < 1$. Since $\varphi'(x) = (1-\alpha)[(x+h)^{-\alpha} - x^{-\alpha}] < 0$, the function φ monotonely decreases for $x > 0$ and fixed $h > 0$. Clearly, $\varphi(x) \leqslant 2^{1-\alpha} h^{1-\alpha}$ for $0 < x \leqslant h$, which implies by virtue of the monotone decrease of φ that the indicated inequality holds for all $x > 0$.

Suppose now $\Delta x_n = -h, h > 0, x_n > 0$. If $0 < h < x_n$, we have, setting $x_n' = x_n - h$,

$$\begin{aligned} |\Delta_{x_n}^{(1)}(f, h)| &= |f(x_1, \ldots, x_{n-1}, x_n - h) - f(x_1, \ldots, x_n)| \\ &= |f(x_1, \ldots, x_{n-1}, x_n')| - f(x_1, \ldots, x_{n-1}, x_n' + h)|, \end{aligned}$$

where $x_n' > 0$ and $h > 0$; i.e. this case reduces to the preceding one. But if $0 < x_n < h$, we have

$$\begin{aligned} |\Delta_{x_n}^{(1)}(f, h)| \leqslant &|f(x_1, \ldots, x_{n-1}, x_n - h) - f(x_1, \ldots, x_{n-1}, 0)| \\ &+ |f(x_1, \ldots, x_n) - f(x_1, \ldots, x_{n-1}, 0)|. \end{aligned}$$

Both of the differences in the right side are estimated in the same way. Let us estimate the second, for example. We first note that the limit

$$\lim_{x_n \to 0} f(x_1, \ldots, x_n) = f(x_1, \ldots, x_{n-1}, 0)$$

exists and is finite for almost all $(x_1, \cdots, x_{n-1})$ provided $\|r^{\alpha} f_{x_n}\|_{\infty}^{(n)} < \infty$. From the definition of a weak derivative it follows that when $x_n > 0$ (analogously when $x_n < 0$ with a replaced by $-a$) we have

$$f(x_1, \ldots, x_{n-1}, x_n) = f(x_1, \ldots, x_{n-1}, a) - \int_{x_n}^{a} f_{x_n}(x_1, \ldots, x_{n-1}, t)\, dt,$$

for almost all $(x_1, \cdots, x_{n-1})$. But

$$\int_{x_n}^{a} |f_{x_n}(x_1, \ldots, x_{n-1}, t)\, dt \leqslant \int_{x_n}^{a} r_{(t)}^{\alpha} |f_{x_n}(x_1, \ldots, x_{n-1}, t)| \frac{dt}{t^{\alpha}}$$

$$\leqslant \| r^{\alpha} f_{x_n} \|_{\infty}^{(n)}{}_{\Gamma_a^{(n)}} \int_0^a \frac{dt}{t^{\alpha}} < \infty.$$

Therefore $|\int_0^a f_{x_n}(x_1, \cdots, x_n) dx_n| < \infty$, which implies that the limit

$$\lim_{x_n \to 0} f(x_1, \ldots, x_n) = f(x_1, \ldots, x_{n-1}, a) - \int_0^a f_{x_n}(x_1, \ldots, x_{n-1}, t)\, dt = f(x_1, \ldots, x_{n-1}, 0)$$

is a finite quantity for the indicated $(x_1, \cdots, x_{n-1})$. Now

$$|f(x_1, \ldots, x_n) - f(x_1, \ldots, x_{n-1}, 0)| = \left| \int_0^{x_n} f_{x_n}(x_1, \ldots, x_{n-1}, t)\, dt \right.$$

$$\leqslant \int_0^h r_{(t)}^{\alpha} |f_{x_n}(x_1, \ldots, x_{n-1}, t)| \frac{dt}{t^{\alpha}} \leqslant \frac{1}{1-\alpha} \|r^{\alpha} f_{x_n}\|_{\infty}^{(n)}{}_{\Gamma_a^{(n)}} h^{1-\alpha}$$

Q.E.D.

PROOF OF (11.34). We write

$$|\Delta_{x_1}^{(1)}(f, h)| \leqslant \Delta_1 + \Delta_2 + \Delta_3,$$

where, using (11.33), we get for Δ_1 and Δ_3

$$\Delta_1 = |f(x_1 + h, x_2, \ldots, x_{n-1}, x_n + h) - f(x_1 + h, x_2, \ldots, x_{n-1}, x_n)|$$

$$\leqslant c_3 \| r^{\alpha} f_{x_n} \|_{p}^{(n)}{}_{\Gamma_a^{(n)}} |h|^{1-\alpha},$$

$$\Delta_3 = |f(x_1, \ldots, x_{n-1}, x_n + h) - f(x_1, \ldots, x_n)| \leqslant c_3 \| r^{\alpha} f_{x_n} \|_{\infty}^{(n)}{}_{\Gamma_a^{(n)}} |h|^{1-\alpha}.$$

To estimate Δ_2 we take $x_n > 0$ and $h > 0$ (the remaining cases are analyzed analogously). Then

$$\Delta_2 = |f(x_1 + h, x_2, \ldots, x_{n-1}, x_n + h) - f(x_1, x_2, \ldots, x_{n-1}, x_n + h)|$$

$$= \left| \int_{x_1}^{x_1 + h} f_{x_1}(u, x_2, \ldots, x_{n-1}, x_n + h)\, du \right|.$$

Let $r_h = \sqrt{x_{m+1}^2 + \cdots + x_{n-1}^2 + (x_n + h)^2}$. Then $r_h > x_n + h > h$, and hence

$$\Delta_2 \leqslant \frac{1}{h^\alpha} \int_{x_1}^{x_1+h} r_h^\alpha \, | f_{x_1}(u, x_2, \ldots, x_{n-1}, x_n + h) | \, du \leqslant \| r^\alpha f_{x_1} \|_{\infty, \Gamma_a^{(n)}}^{(n)} h^{1-\alpha}.$$

Thus

$$\| \Delta_{x_1}^{(1)}(f, h) \|_{\infty, \Gamma_{\eta\xi}^{(n)}}^{(n)} \leqslant (2c_3 \| r^\alpha f_{x_n} \|_{\infty, \Gamma_a^{(n)}}^{(n)} + \| r^\alpha f_{x_1} \|_{\infty, \Gamma_a^{(n)}}^{(n)}) | h |^{1-\alpha},$$

Q.E.D.

[11.9]. *Suppose* $\sigma \geqslant 0$, *a function* f *is defined on the domain* $\Gamma_a^{(n)}$, *all of its weak derivatives up to order* $[\sigma] + 1$ *inclusively*(4) *exist on the set* $\Gamma_a^{(n)} \setminus E^m$ *and*

$$\| r^\sigma f_{x_1^{\sigma_1} \ldots x_n^{\sigma_n}}^{([\sigma]+1)} \|_{\infty, \Gamma_a^{(n)}}^{(n)} < \infty \quad \text{for all} \quad \sigma_1 + \ldots + \sigma_n = [\sigma] + 1.$$

Then for any $\epsilon \in (0, 1 - \sigma + [\sigma])$

$$\| r^{\sigma - t - 1 + \frac{t+1}{[\sigma]}\epsilon} f_{x_1^{t_1} \ldots x_n^{t_n}}^{[\sigma]-t} \|_{\infty, \Gamma_a^{(n)}}^{(n)} \leqslant M_t, \quad t = 0, 1, \ldots, [\sigma] - 1, \; t_1 + \ldots + t_n = [\sigma] - t,$$

$$\| f \|_{\infty, \Gamma_a^{(n)}}^{(n)} \leqslant M_{[\sigma]},$$

where

$$M_t \leqslant C_t \sum_{l=0}^{t} \sum_{l_1 + \ldots + l_n = [\sigma] - l} \left(\| f_{x_1^{l_1} \ldots x_n^{l_n}}^{[\sigma]-l} \|_{\infty, \Gamma_{-an}^{(n-1)}}^{(n-1)} + \| f_{x_1^{l_1} \ldots x_n^{l_n}}^{[\sigma]-l} \|_{\infty, \Gamma_{an}^{(n-1)}}^{(n-1)} \right)$$
$$+ D_t \sum_{\sigma_1 + \ldots + \sigma_n = [\sigma]+1} \| r^\sigma f_{x_1^{\sigma_1} \ldots x_n^{\sigma_n}}^{[\sigma]+1} \|_{\infty, \Gamma_a^{(n)}}^{(n)},$$

(4) $[\sigma]$ denotes, as usual, the integral part of σ.

the constants C_t *and* D_t *not depending on* f. *Finally,*

$$f \in H^{1+[\sigma]-\sigma+\varepsilon}_{\infty(n)}(M_{[\sigma]}, \Gamma^{(n)}_a).$$

This theorem follows from inequalities (11.31)–(11.34) in the same way as Theorem [11.7] follows from Theorems [11.1]–[11.6].

We conclude this section by giving a criterion for the extendability of functions and obtaining several more integral inequalities, which we need for the sequel and whose methods of proof closely adjoin the methods of the present section.

[11.10]. *Suppose* $0 \leqslant \alpha < p-1, m = n-1, r = |x_n|, f(x_1, \cdots, x_{n-1}, 0) = 0$ (*see the Corollary to Theorem* [11.3]) *and* $|h| < a$. *Then there exists a constant* $A > 0$ *depending only on* p *and* a *such that*

$$\left\{\int_\Gamma \cdots \int dx_1 \ldots dx_{n-1} \int_0^h |x_n|^\alpha |f(x_1, \ldots, x_n)|^p \, dx_n\right\}^{\frac{1}{p}} \leqslant A \, \| f_{x_n} \|^{(n)}_{p(|x_n|, \alpha) \, \Gamma^{(n)}_a} |h| \, (5)$$

PROOF. If $\|f_{x_n}\|^{(n)}_{p(|x_n|, \alpha)} \geqslant \infty$, the inequality is trivial. If $\|f_{x_n}\|^{(n)}_{p(|x_n|, \alpha)} < \infty$, Theorem [11.3] implies that for almost all $(x_1, \cdots, x_{n-1})$ the derivative f_{x_n} is summable with respect to x_n over segments lying in $\Gamma^{(n)}_a$, and hence

$$f(x_1, \ldots, x_n) = \int_0^{x_n} f_{x_n}(x_1, \ldots, x_{n-1}, t) \, dt.$$

Consider the case $h > 0$ (the proof is analogous for $h < 0$). Applying Hölder's inequality, we get

(5) As can be seen from the following proof, the left side of this inequality is actually $O(h)$; but this information will not be needed in the sequel.

$$\left\{\int_\Gamma \cdots \int\int^h x_n^\alpha |f(x_1, \ldots, x_n)| \; dx_1 \ldots dx_n\right\}^{\frac{1}{p}}$$

$$= \left\{\int_\Gamma \cdots \int\int_0^h x_n^\alpha \left|\int_0^{x_n} f_{x_n}(x_1, \ldots, x_{n-1}, t)\, dt\right|^p dx_1 \ldots dx_n\right\}^{\frac{1}{p}}$$

$$\leqslant \left\{\int_\Gamma \cdots \int\int_0^h x_n^\alpha \int_0^{x_n} t^\alpha |f_{x_n}(x_1 \ldots, x_{n-1}, t)|^p dt \left(\int_0^{x_n} \frac{dt}{t^{\frac{\alpha q}{p}}}\right)^{\frac{p}{q}} dx_1 \ldots dx_n\right\}^{\frac{1}{p}}$$

$$\leqslant \left(\frac{p-1}{p-1-\alpha}\right)^{\frac{1}{q}} \left\{\int_0^h x_n^{p-1} dx_n \int_\Gamma \cdots \int\int_0^a t^\alpha |f_{x_n}(x_1, \ldots, x_{n-1}, t)|^p dt\right\}^{\frac{1}{p}}$$

$$\leqslant \left(\frac{p-1}{p-1-\alpha}\right)^{\frac{1}{q}} \frac{1}{p^{\frac{1}{p}}} \|f_{x_n}\|^{(n)}_{p(|x_n|, \alpha)\, \Gamma^{(n)}_{+an}} h.$$

Q.E.D.

[11.11]. *Suppose* $0 \leqslant \alpha < p-1, m = n-1$ *and*

$$\|f_{x_n}\|^{(n)}_{p(x_n, \alpha)\, \Gamma^{(n)}_{+an}} < \infty, \|f\|^{(n)}_{p\, \Gamma^{(n)}_{+an}} < \infty.$$

Then there exists a constant $B > 0$ *not depending on* f *such that*

$$\left| \|f\|^{(n-1)}_{p\, \Gamma} - \|f\|^{(n-1)}_{p\, \Gamma^{(n-1)}_{+an}} \right| \leqslant B a^{\frac{p-1-\alpha}{p}} \|f_{x_n}\|^{(n)}_{p(x_n, \alpha)\, \Gamma^{(n)}_{+an}}, \tag{11.35}$$

where the value of f *on the* $(n-1)$- *dimensional domain* Γ *is understood in the sense of a mean value while the constant* $B > 0$ *does not depend on* f *or* a.

PROOF. As we already know, it follows from the assumptions of the theorem that (i) $f(x_1, \cdots, x_{n-1}, a)$ exists in the sense of convergence in the mean and of $\|f\|^{(n-1)}_{p\, \Gamma^{(n-1)}_{+an}}$ (see [11.6]), and (ii) the limit $\lim_{x_n \to 0} f(x_1, \cdots, x_{n-1}, x_n)$ exists and is finite for almost all $(x_1, \cdots, x_{n-1})$ (see [11.3]). We show that it is also a limit in the mean:

$$\int_\Gamma \cdots \int |f(x_1, \ldots x_{n-1}, x_n) - f(x_1, \ldots, x_{n-1}, 0)|^p\, dx_1 \ldots dx_{n-1}$$

$$= \int_\Gamma \cdots \int \left| \int_0^{x_n} f_{x_n}(x_1, \ldots, x_{n-1}, t)\, dt \right|^p dx_1 \ldots dx_{n-1}$$

$$\leqslant \int_\Gamma \cdots \int \int_0^{x_n} t^\alpha |f_{x_n}(x_1 \ldots, x_{n-1}, t)|^p\, dt \left(\int_0^{x_n} \frac{dt}{t^{\frac{\alpha q}{p}}} \right)^{\frac{p}{q}} dx_1 \ldots dx_n$$

$$\leqslant \left(\frac{p-1}{p-1-\alpha} \right)^{p-1} \| f_{x_n} \|^{(n)}_{\substack{p(x_n, \alpha) \\ \Gamma^{(n)}_{+an}}} x_n^{p-1-\alpha} \to 0$$

for $x_n \to 0$. To prove inequality (11.35) we note that

$$\| f \|^{(n-1)}_{\substack{p \\ \Gamma}} = \left\{ \int_\Gamma \cdots \int |f(x_1, \ldots, x_{n-1}, 0)|^p\, dx_1 \ldots dx_{n-1} \right\}^{\frac{1}{p}}$$

$$= \left\{ \int_\Gamma \cdots \int \left| f(x_1, \ldots, x_{n-1}, a) - \int_0^a f_{x_n}(x_1, \ldots, x_{n-1}, t)\, dt \right|^p dx_1 \ldots dx_{n-1} \right\}^{\frac{1}{p}}$$

$$\leqslant \left\{ \int_\Gamma \cdots \int |f(x_1, \ldots, x_{n-1}, a)|^p\, dx_1 \ldots dx_{n-1} \right\}^{\frac{1}{p}}$$

$$+ \left\{ \int_\Gamma \cdots \int \left| \int_0^a f_{x_n}(x_1, \ldots, x_{n-1}, t)\, dt \right|^p dx_1 \ldots dx_{n-1} \right\}^{\frac{1}{p}}$$

$$\leqslant \| f \|^{(n-1)}_{\substack{p \\ \Gamma^{(n-1)}_{+an}}} + \left\{ \int_\Gamma \cdots \int \int_0^a t^\alpha |f_{x_n}(x_1, \ldots, x_{n-1}, t)|^p\, dt \left(\int_0^a \frac{dt}{t^{\frac{\alpha q}{p}}} \right)^{\frac{p}{q}} dx_1 \ldots dx_{n-1} \right\}^{\frac{1}{p}}$$

$$\leqslant \| f \|^{(n-1)}_{\substack{p \\ \Gamma^{(n-1)}_{+an}}} + \left(\frac{p-1}{p-1-\alpha} \right)^{\frac{1}{q}} a^{\frac{p-1-\alpha}{p}} \| f_{x_n} \|^{(n)}_{\substack{p(x_n, \alpha) \\ \Gamma^{(n)}_{+an}}}.$$

The other part of inequality (11.35) is proved analogously.

[11.12]. *Suppose* $\alpha \geqslant p-1$, $m = n-1$, $\operatorname{mes} \Gamma < \infty$, $\| f_{x_n} \|^{(n)}_{\substack{p(x_n, \alpha) \\ \Gamma^{(n)}_{+an}}} < \infty$, $\| f \|^{(n-1)}_{\substack{p \\ \Gamma^{(n-1)}_{+an}}} < \infty$, $\epsilon > 0$, $f_\epsilon(x_1, \cdots, x_n) = x_n^{\alpha/n - 1/q + \epsilon} f(x_1, \cdots, x_n)$. *Then there exists a constant* $c > 0$ *not depending on* f *such that*

$$\int_{\Gamma^{(n)}_{+an}} \cdots \int \left| \frac{\partial}{\partial x_n} f_\epsilon(x_1 \ldots x_n) \right| dx_1 \ldots dx_n \leqslant c \left[\| f \|^{(n)}_{\substack{p\left(x_n, \alpha - p + \frac{p\epsilon}{2}\right) \\ \Gamma^{(n)}_{+an}}} + \| f_{x_n} \|^{(n)}_{\substack{p(x_n, \alpha) \\ \Gamma^{(n)}_{+an}}} \right] < \infty. \tag{11.36}$$

COROLLARY. *For almost all* $(x_1, \cdots, x_{n-1})$ *the function* $f_\epsilon(x_1, \cdots, x_n)$

is absolutely continuous in x_n *on the segment* $[0, a]$, $f_\epsilon(x_1, \cdots, x_{n-1}, 0) = 0$, *and hence*

$$f_\varepsilon(x_1 \ldots, x_n) = \int_0^{x_n} \frac{\partial f_\varepsilon(x_1, \ldots, x_{n-1}, t)}{\partial t} dt.$$

PROOF. We first note that the condition $\alpha \geqslant p - 1$ implies $\alpha p^{-1} - q^{-1} + \epsilon > \epsilon > 0$. We now have

$$\int \ldots \int_{\Gamma^{(n)}_{+an}} \left| \frac{\partial}{\partial x_n} f_\varepsilon(x_1, \ldots, x_n) \right| dx_1 \ldots dx_n = \int \ldots \int_{\Gamma^{(n)}_{+an}} \left| \frac{\partial}{\partial x_n} \left[x_n^{\frac{\alpha}{p} - \frac{1}{q} + \varepsilon} \right.\right.$$

$$\times f(x_1, \ldots x_n) \Big| dx_1 \ldots dx_n \leqslant \left(\frac{\alpha}{p} - \frac{1}{q} + \varepsilon\right) \int \ldots \int_{\Gamma^{(n)}_{+an}} x_n^{\frac{\alpha}{p} - 1 + \frac{\varepsilon}{2}} f(x_1, \ldots, x_n)$$

$$\times \frac{dx_1 \ldots dx_n}{x_n^{\frac{1}{q} - \frac{\varepsilon}{2}}} + \int \ldots \int_{\Gamma^{(n)}_{+an}} x_n^{\frac{\alpha}{p}} | f_{x_n}(x_1, \ldots, x_n) | \frac{dx_1 \ldots dx_n}{x^{\frac{1}{q} - \varepsilon}}$$

$$\leqslant \left(\frac{\alpha}{p} - \frac{1}{q} + \varepsilon\right) \left\{ \int \ldots \int_{\Gamma^{(n)}_{+an}} x_n^{\alpha - p + \frac{p\varepsilon}{2}} | f(x_1, \ldots, x_n) |^p dx_1 \ldots dx_n \right\}^{\frac{1}{p}}$$

$$\times \left\{ \int \ldots \int_{\Gamma} dx_1 \ldots dx_{n-1} \int_0^a \frac{dx_n}{x_n^{1 - \frac{\varepsilon q}{2}}} \right\}^{\frac{1}{q}} + \left\{ \int \ldots \int_{\Gamma^{(n)}_{+an}} x_n^{\alpha} | f_{x_n}(x_1, \ldots, x_n) |^p dx_1 \ldots dx_n \right\}^{\frac{1}{p}}$$

$$\times \left\{ \int \ldots \int_{\Gamma} dx_1 \ldots dx_{n-1} \int_0^a \frac{dx_n}{x_n^{1-q\varepsilon}} \right\}^{\frac{1}{q}} \leqslant \left(\frac{\alpha}{p} - \frac{1}{q} + \varepsilon\right) \left(\frac{2}{q\varepsilon}\right)^{\frac{1}{q}} a^{\frac{\varepsilon}{2}} (\text{mes}\, \Gamma)^{\frac{1}{q}} \| f \|^{(n)}_{\substack{p\,(x_n,\, \alpha - p + \frac{p\varepsilon}{2}) \\ \Gamma^{(n)}_{+an}}}$$

$$+ \left(\frac{1}{q\varepsilon}\right)^{\frac{1}{q}} a^{\varepsilon} (\text{mes}\, \Gamma)^{\frac{1}{q}} \| f_{x_n} \|^{(n)}_{\substack{p\,(x_n,\, \alpha) \\ \Gamma^{(n)}_{+an}}}.$$

Inequality (11.36) is proved. By virtue of Theorem [11.1] the finiteness of $\|f_{x_n}\|^{(n)}_{p(x_n,a), \Gamma^{(n)}_{+an}}$ and $\|f\|^{(n-1)}_{p, \Gamma^{(n-1)}_{+an}}$ implies the finiteness of $\|f\|^{(n)}_{p(x_n, \alpha - p + p\epsilon/2), \Gamma^{(n)}_{+an}}$. Therefore, under the assumptions of the theorem, $\int \cdots \int_{\Gamma^{(n)}_{+an}} |\partial f_\epsilon / \partial x_n| dx_1 \cdots dx_n < \infty$, and hence by virtue of Fubini's theorem the integral $\int_0^a |\partial f_\epsilon/\partial x_n| dx_n$ is finite for almost all $(x_1, \cdots, x_{n-1}) \in \Gamma$. It follows that the limit

$$\lim_{x_n \to 0} f_\varepsilon(x_1, \ldots, x_n) = \lim_{x_n \to 0} \left[f_\varepsilon(x_1, \ldots, x_{n-1}, a) - \int_{x_n}^{a} \frac{\partial f_\varepsilon}{\partial x_n} dt \right] = f_\varepsilon(x_1, \ldots, x_{n-1}, 0)$$

exists and is finite. Noting, further, that

$$f_\varepsilon(x_1, \ldots x_n) = x_n^{\frac{\varepsilon}{2}} f_{\frac{\varepsilon}{2}}(x_1, \ldots, x_n)$$

and that the function $f_{\epsilon/2}$ is bounded for admissible values of $(x_1, \cdots, x_{n-1})$, we get $f_\epsilon(x_1, \cdots, x_{n-1}, 0) = 0$ for almost all $(x_1, \cdots, x_{n-1}) \in \Gamma$.

[11.13]. *Suppose* $\alpha \geqslant p-1$, $m = n-1$, $0 < h < \xi$, $\epsilon > 0$, $\|f\|_{p}^{(n-1)} < \infty$, $\Gamma_{+an}^{(n-1)}$

$\|f_{x_n}\|_{p(x_n,\alpha)}^{(n)} < \infty$. *Then there exists a constant* $c' > 0$ *not depending on* f *such* $\Gamma_{+an}^{(n)}$

that

$$\left\{ \int_\Gamma \cdots \int dx_1 \ldots dx_{n-1} \int_0^h x_n^{\alpha+\varepsilon} |f(x_1, \ldots, x_n)|^p dx_n \right\}^{\frac{1}{p}} \tag{11.37}$$

$$\leqslant c' \Big[\|f\|^{(n)}_{\substack{p\left(x_n,\, \alpha-p+\frac{\varepsilon}{2}\right)\\ \Gamma^{(n)}_{+an}}} + \|f_{x_n}\|^{(n)}_{\substack{p(x_n,\, \alpha)\\ \Gamma^{(n)}_{+an}}} \Big] h.$$

PROOF. Let

$$f_\varepsilon(x_1, \ldots, x_n) = x_n^{\frac{\alpha}{p} - \frac{1}{q} + \frac{\varepsilon}{p}} f(x_1, \ldots, x_n).$$

Then, applying the Corollary to the preceding theorem and Hölder's inequality, we get

$$\left\{ \int_\Gamma \cdots \int \int_0^h x_n^{\alpha+\varepsilon} |f(x_1, \ldots, x_n)|^p \, dx_1 \ldots dx_n \right\}^{\frac{1}{p}}$$

$$= \left\{ \int_\Gamma \cdots \int \int_0^h |f_\varepsilon(x_1, \ldots x_n)|^p x_n^{\frac{p}{q}} dx_1 \ldots dx_n \right\}^{\frac{1}{p}} =$$

$$= \left\{ \int_\Gamma \cdots \int \int_0^h \left| \int_0^{x_n} \frac{\partial f_\varepsilon (x_1, \ldots, x_{n-1}, t)}{\partial t} dt \right|^p x_n^{\frac{p}{q}} dx_1 \ldots dx_n \right\}^{\frac{1}{p}}$$

$$\leqslant \left(\frac{\alpha}{p} - \frac{1}{q} + \frac{\varepsilon}{p}\right) \left\{ \int_\Gamma \cdots \int \int_0^h x_n^{\frac{p}{q}} \left| \int_0^{x_n} t^{\frac{\alpha}{p} - \frac{1}{q} - 1 + \frac{\varepsilon}{p}} f(x_1, \ldots, x_{n-1}, t)\, dt \right|^p dx_1 \ldots dx_n \right\}^{\frac{1}{p}}$$

$$+ \left\{ \int_\Gamma \cdots \int \int_0^h x_n^{\frac{p}{q}} \left| \int_0^{x_n} t^{\frac{\alpha}{p} - \frac{1}{q} + \frac{\varepsilon}{p}} f_{x_n}(x_1, \ldots, x_{n-1}, t)\, dt \right|^p dx_1 \ldots dx_n \right\}^{\frac{1}{p}}$$

$$\leqslant \left(\frac{\alpha}{p} - \frac{1}{q} + \frac{\varepsilon}{p}\right) \left\{ \int_\Gamma \cdots \int \int_0^h x_n^{p-1} \left[\int_0^{x_n} t^{\alpha - p + \frac{\varepsilon}{2}} \left| f(x_1, \ldots, x_{n-1}, t) \right|^p dt \right] \right.$$

$$\times \left. \left[\int_0^{x_n} \frac{dt}{t^{1 - \frac{\varepsilon q}{2p}}} \right]^{\frac{p}{q}} dx_1 \ldots dx_n \right\}^{\frac{1}{p}} + \left\{ \int_\Gamma \cdots \int \int_0^h x_n^{p-1} \left[\int_0^{x_n} t^\alpha \left| f_{x_n}(x_1, \ldots, x_{n-1}, t) \right|^p dt \right] \right.$$

$$\times \left. \left[\int_0^{x_n} \frac{dt}{t^{1 - \frac{\varepsilon q}{p}}} \right]^{\frac{p}{q}} dx_1 \ldots dx_n \right\}^{\frac{1}{p}} \leqslant \left(\frac{\alpha}{p} - \frac{1}{q} + \frac{\varepsilon}{p}\right) \left(\frac{2p}{\varepsilon q}\right)^{\frac{1}{q}}$$

$$\times \left\{ \int_0^h x_n^{p - 1 - \frac{\varepsilon}{2}} dx_n \int_\Gamma \cdots \int \int_0^a t^{\alpha - p + \frac{\varepsilon}{2}} \left| f(x_1, \ldots, x_{n-1}, t) \right|^p dx_1 \ldots dx_{n-1} dt \right\}^{\frac{1}{p}}$$

$$+ \left(\frac{p}{\varepsilon q}\right)^{\frac{1}{q}} \left\{ \int_0^h x_n^{p-1+\varepsilon} dx_n \int_\Gamma \cdots \int \int_0^a t^\alpha \left| f_{x_n}(x_1, \ldots x_{n-1}, t) \right|^p dx_1 \ldots dx_{n-1} dt \right\}^{\frac{1}{p}}$$

$$\leqslant \left[\left(\frac{\alpha}{p} - \frac{1}{q} + \frac{\varepsilon}{p}\right) \left(\frac{2p}{\varepsilon q}\right)^{\frac{1}{q}} \frac{1}{(p + \varepsilon)^{\frac{1}{p}}} \| f \|^{(n)}_{\Gamma^{(n)}_{+an} p \left(x_n, \alpha - p + \frac{\varepsilon}{2}\right)} \right.$$

$$\left. + \left(\frac{p}{\varepsilon q}\right)^{\frac{1}{q}} \frac{1}{(p + \varepsilon)^{\frac{1}{p}}} \| f_{x_n} \|^{(n)}_{\Gamma^{(n)}_{+an} p (x_n, \alpha)} \right] h.$$

Q.E.D.

[11.14]. *Suppose a function* f *is defined on the domain* $\Gamma^{(n)}_{+an}$ *that together with all of its weak derivatives up to order* $\bar{r}$ *inclusively is* p *power summable on* $\Gamma^{(n)}_{+an}$, $\alpha \neq p - 1$, $r = \bar{r} + \alpha$, $m = n - 1$ *and*

$$\| f^{(\bar{r}+1)}_{x_1^{r_1} \ldots x_n^{r_n}} \|^{(n)}_{\Gamma^{(n)}_{+an} p (x_n, p(1-\alpha))} < \infty$$

for all $r_1 + \cdots + r_n = \bar{r} + 1$. *Then there exists a function* F *defined on* $\Gamma^{(n)}_a$, *coinciding with* f *on* $\Gamma^{(n)}_{+an}$ *and such that* $F\in H^{(r)}_{p(n)}(M,\Gamma^{(n)}_a)$, *where*

$$M\leqslant A_1\sum_{r_1+\dots+r_n=\bar r+1}\|f^{(\bar r+1)}_{x_1^{r_1}\dots x_n^{r_n}}\|^{(n)}_{\substack{p(x_n,\ p(1-\alpha))\\ \Gamma^{(n)}_{+an}}}+A_2\sum_{\rho_1+\dots+\rho_n=\bar r}\|f^{(\bar r)}_{x_1^{\rho_1}\dots x_n^{\rho_n}}\|^{(n-1)}_{\substack{p\\ \Gamma^{(n-1)}_{+an}}},\tag{11.38}$$

the constants A_1 *and* A_2 *not depending on* f.

Proof. We take the function F constructed in the proof of Theorem [6.2] (see (6.7)) and estimate the norms of its derivatives of order $\bar r+1$:

$$\|F^{(\bar r+1)}_{x_1^{r_1}\dots x_n^{r_n}}\|^{(n)}_{\substack{p(x_n,\ p(1-\alpha))\\ \Gamma^{(n)}_{-an}}}$$

$$=\left\{\int\limits_{\Gamma}\cdots\int\int\limits_{-a}^{0}|x_n|^{p(1-\alpha)}\left|\sum_{k=1}^{\bar r+1}\lambda_k\left(-\frac{1}{k}\right)^{r_n}f^{(\bar r+1)}_{x_1^{r_1}\dots x_n^{r_n}}\left(x_1,\dots,x_{n-1},-\frac{x_n}{k}\right)\right|^{p}dx_1\dots dx_n\right\}^{\frac{1}{p}}$$

$$\leqslant\sum_{k=1}^{\bar r+1}|\lambda_k|\left\{\int\limits_{\Gamma}\cdots\int\int\limits_{0}^{\frac{a}{k}}(kt)^{p(1-\alpha)}|f^{(\bar r+1)}_{x_1^{r_1}\dots x_n^{r_n}}(x_1,\dots,x_{n-1},t)|^{p}dx_1\dots dx_{n-1}dt\right\}^{\frac{1}{p}}$$

$$\leqslant\|f^{(\bar r+1)}_{x_1^{r_1}\dots x_n^{r_n}}\|^{(n)}_{\substack{p(x_n,\ p(1-\alpha))\\ \Gamma^{(n)}_{+an}}}\sum_{k=1}^{\bar r+1}|\lambda_k|(\bar r+1)^{\frac{1}{p}}.\tag{11.39}$$

If $\alpha=1$, we have

$$\|f^{(\bar r+1)}_{x_1^{r_1}\dots x_n^{r_n}}\|^{(n)}_{\substack{p\\ \Gamma^{(n)}_a}}<\infty,$$

which implies by virtue of (11.39) that

$$\|F^{(\bar r+1)}_{x_1^{r_1}\dots x_n^{r_n}}\|^{n}_{\substack{p\\ \Gamma^{(n)}_{+an}}}<\infty,$$

and hence the assertion of the throrem is trivial. And if $\alpha<1$, our theorem immediately follows by virtue of the same estimate (11.39) from the imbedding theorem [11.7] if one applies it to the functions $F^{(\bar r)}_{x_1^{\bar r}},\dots,F^{(\bar r)}_{x_n^{\bar r}}$ (for $s=0$).

Finally, we note that in the case $\alpha=p-1$ the same Theorem [11.7] implies that $F\in H_p^{(r-\epsilon)}(M)$ for any $\epsilon>0$ under preservation of all of the other assertions of the theorem.

§12. Weighted imbedding theorems for function classes on arbitrary manifolds

We will use the terminology of §8 in the present section.

[12.1]. *Suppose* $1 \leqslant p < \infty, \Gamma$ *is a submanifold of an m-dimensional manifold* $K^{(m)}$ *of order of smoothness* $k \geqslant 2, \overline{\Gamma}$ *is compact and* $\overline{\Gamma} \subset K^{(m)}$. *Suppose further,* G *is a domain in* E^n, $G \supset \overline{\Gamma}$, *a function* f *together with its weak derivatives up to order* $s + 1$ *inclusively is defined on the set* $G \setminus K^{(m)}$, $s \leqslant k - 2$ *and for some* $\xi > 0$

$$\| f^{(s+1)}_{x_1^{s_1} \ldots x_n^{s_n}} \|^{(n)}_{\substack{p(r,\sigma) \\ \Gamma_\xi^{(n)}}} < \infty$$

for all $s_1 + \cdots + s_n = s + 1$, *where* $\overline{\Gamma}_\xi^{(n)} \setminus \overline{\Gamma} \subset G \setminus K^{(m)}$, $\sigma = sp + \alpha$, s *is a nonnegative integer and* $0 \leqslant \alpha < p$. *Then for any* $\epsilon \in (0, p - \alpha)$

$$\| f^{(s-t)}_{x_1^{t_1} \ldots x_n^{t_n}} \|^{(n)}_{\substack{p \\ \Gamma_\xi^{(n)}}} \left[r, \sigma - (t+1)p + \frac{t+1}{2s} \varepsilon \right] \leqslant M_t, \; t = 0, 1, \ldots, s - 1, \; t_1 + \ldots + t_n = s - t,$$

$$\| f \|^{(n)}_{\substack{p \\ \Gamma_\xi^{(n)}}} \leqslant M_s,$$

$$M_t \leqslant A_t \sum_{l=0}^{t} \sum_{l_1 + \ldots + l_n = s - l} \left[\| f^{(s-l)}_{x_1^{l_1} \ldots x_n^{l_n}} \|^{(n-1)}_{\substack{p \\ \Gamma^{(n-1)}_{+\xi n}}} + \| f^{(s-l)}_{x_1^{l_1} \ldots x_n^{l_n}} \|^{(n-1)}_{\substack{p \\ \Gamma^{(n-1)}_{-\xi n}}} \right]$$

$$+ B_t \sum_{s_1 + \ldots + s_n = s+1} \| f^{(s+1)}_{x_1^{s_1} \ldots x_n^{s_n}} \|^{(n)}_{\substack{p(r,\sigma) \\ \Gamma_\xi^{(n)}}}, \quad t = 0, 1, \ldots s,$$

the constants A_t *and* B_t *not depending on* f. *Finally,*

$$f \in H^{\left(1 - \frac{\alpha + \varepsilon}{p}\right)}_{p(n)} (M_s, \Gamma_\xi^{(n)})$$

(ϵ *can be set equal to zero when* $s = 0$ *and* $\alpha \neq p - 1$).

In the case $p = \infty$ *one should put* $s = [\sigma]$ *and, in all of the exponents of the distance* r, *take* $p = 1$.

PROOF. By virtue of the definition of the cylindroid $\Gamma_\xi^{(n)}$ (see (8.8)) there exist a finite number of canonical neighborhoods (see (8.5)) $U^{(\nu)}$ and $V^{(\nu)}$ ($\nu = 1, \cdots, \nu_0$) of heights ξ and a respectively, $0 < \xi < a$, corresponding to points $P \in \overline{\Gamma}$ and such that

$$\Gamma_\xi^{(n)} \subset \bigcup_{\nu=1}^{\nu_0} U^{(\nu)}, \quad \bar{U}^{(\nu)} \subset V^{(\nu)}, \quad \nu = 1, 2, \ldots, \nu_0.$$

We denote by $x_{(\nu)}(u)$ the canonical mapping (see (8.3)) of the domain $V_u^{(\nu)}$ onto $V^{(\nu)}$. By virtue of the conditions of the theorem it is at least $s+1$ times boundedly differentiable. Then for each $\nu = 1, \cdots, \nu_0$ the function $f(x_{(\nu)}(u))$ satisfies the conditions of Theorem [11.7] (Theorem [11.9] when $p=\infty$) on $V_u^{(\nu)}$ if one assumes that the set $V_u^{(\nu)}$ corresponds to a domain $\Gamma_a^{(n)}$ in the notation of §11. Therefore all of the assertions of our theorem are valid to within constants for the case $U^{(\nu)} = \Gamma_\xi^{(n)}$, a change in the constants of Theorem [11.7] (Theorem [11.9]) being possible only at the expense of the properties of the canonical mappings $x_{(\nu)}(u)$ and hence not depending on the function f. It follows by virtue of the finiteness of the number of canonical neighborhoods $U^{(\nu)}$ that Theorem [12.1] is proved.

[12.2]. *Suppose G is a finite domain in E^n that is bounded by an $(n-1)$-dimensional manifold K of smoothness $k \geqslant 2$ (or by a finite number of such manifolds) on which a consistent choice of normals has been fixed so that the direction of the inward normal is taken as the positive direction, $K_{+\eta}^{(n)} = K_{+\eta}^{(n)}(K)$ and $K_{+\eta}^{(n-1)} = K_{+\eta}^{(n-1)}(K)$. Suppose, further, there is defined on G a function $f \in H_{p(n)}^{(r)}(M, G)$, $r = \bar{r} + \alpha$, $\alpha \neq p-1$, $\bar{r}+1 \leqslant k-1$, there exists an $\eta > 0$ such that all of the weak derivatives of order $\bar{r}+1$ of f exist on $K_{+\eta}^{(n)}$, and*

$$\left\| f_{x_1^{r_1} \ldots x_n^{r_n}}^{(\bar{r}+1)} \right\|_{p(r,\, p(1-\alpha))}^{(n)} < \infty, \quad r_1 + \ldots + r_n = \bar{r} + 1.$$

Then $f \in H_{p(n)}^{(r)}(M^*, G, E^n)$, where

$$M^* \leqslant d \left[\sum_{\substack{r_1+\ldots+r_n=\\=\bar{r}+1}} \left\| f_{x_1^{r_1} \ldots x_n^{r_n}}^{(\bar{r}+1)} \right\|_{\substack{p(r,\, p(1-\alpha)) \\ K_{+\eta}^{(n)}}}^{(n)} + \sum_{\rho_1+\ldots+\rho_n=\bar{r}} \left\| f_{x_1^{\rho_1} \ldots x_n^{\rho_n}}^{(\bar{r})} \right\|_{\substack{p \\ K_{+\eta}^{(n-1)}}}^{(n-1)} + M + \| f \|_{\substack{p \\ G}}^{(n)} \right].$$

The proof of this theorem is carried out by a method analogous to that of the proof of [8.3]. We note that [12.2] can be generalized (analogously to Theorem [8.3]) to the case when G is a submanifold of a manifold; but this will not be done for the sake of simplicity.

Consider a covering of the manifold K by a finite number of the canonical neighborhoods $V^{(\nu)}$ of height $\eta > 0$, and let $x_{(\nu)}(u)$ be the corresponding mappings. Further, suppose $\{U^{(\nu)}\}$ is a covering of K by the canonical neighborhoods

$U^{(\nu)}$ of height $\eta' < \eta$ such that $\bar{U}^{(\nu)} \subset V^{(\nu)}$, $\nu = 1, \cdots, \nu_0$. Then the function $f(x_{(\nu)}(u))$ satisfies all of the conditions of Theorem [11.14] on $\overset{+}{V}_u$ (see (8.7)) and hence can be extended as a function $F_{(\nu)}(x_{(\nu)}(u))$ defined on $V_u^{(\nu)}$ (see (8.5)), all of the assertions of Theorem [11.14] being satisfied for it. We now define on $V^{(\nu)}$ the function $F_{(\nu)}(P) = F_{(\nu)}(x_{(\nu)}(u))$, where $P = x_{(\nu)}(u)$. It has the following properties.

1°. $F_{(\nu)}$ coincides with f on $\overset{+}{V}_\nu$.

2°. $F_{(\nu)} \in H_{p(n)}^{(r)}(M_\nu, U^{(\nu)})$, where

$$M_\nu = d^{(\nu)} \left[\sum_{r_1 + \dots + r_n = \bar{r}+1} \left\| f^{(\bar{r}+1)}_{x_1^{r_1} \dots x_n^{r_n}} \right\|^{(n)}_{p(r,\, p(1-\alpha))\, \overset{+}{V}(\nu)} + \sum_{\rho_1 + \dots + \rho_n = \bar{r}} \left\| f^{(\bar{r})}_{x_1^{\rho_1} \dots x_n^{\rho_n}} \right\|^{(n-1)}_{p\, K^{(n)}_{+\eta}(\overset{\circ}{V}(\nu))} + M + \|f\|^{(n)}_{p\, \overset{+}{V}(\nu)} \right].$$

3°. $\|F_\nu\|^{(n)}_{p\, \overset{+}{V}(\nu)} \leqslant M_\nu$.

This readily follows from theorem [11.14] and the theorem of Nikol'skiĭ on the behavior of the H-classes under differentiable mappings ([4], page 59).

Further, for each point $P \in \bar{G}$ we choose a ball Q_P of radius r_P with center at P is such a way that it is either entirely contained in G or in one of the neighborhoods $U^{(\nu)}$. Let Q'_P be the ball of radius $r_P/3$ with center at P. From the covering $\{Q'_P\}$ of the compact set $\bar{G}$ we select a finite subcovering $\{Q'_{P_i}\}$, $i = 1, \cdots, i_0$, and denote by $f^{(i)}$ a function coinciding with f or one of its extensions F_ν on Q_{P_i} and equal to zero outside this ball. We take, as usual, an infinitely differentiable function $\psi(t)$ on the real line such that $\psi(t) = 0$ for $t \leqslant 1$ and $\psi(t) = 1$ for $t \geqslant 2$. We put $\psi_i(P) = \psi(\rho(P, P_i)/r_{P_i})$, $\Psi^{(i)} = \psi_1 \psi_2 \cdots \psi_{i-1}(1 - \psi_i)$ and

$$F(P) = \sum_{i=1}^{i_0} f^{(i)}(P)\, \Psi^{(i)}(P), \quad P \in E^n.$$

The function F effects the desired extension of f from G onto E^n. This is proved in the same way as was indicated in the analogous cases in the proofs of Theorems [6.1], [8.3] and others.

REMARK. It is not difficult to also verify that the function F satisfies the condition

$$\|F^{(\bar r+1)}_{x_1^{r_1}\ldots x_n^{r_n}}\|^{(n)}_{p(r,\,p(1-\alpha))\,K^{(n)}_{\eta}}<\infty,\quad r_1+\cdots+r_n=\bar r+1.$$

Thus when $\alpha = 1$ Theorem [12.2] contains an assertion on the extendability of the Sobolev classes $W_p^{(\bar r+1)}$ ([2], page 45).

[12.3] (IMBEDDING AND EXTENSION THEOREM). *Suppose* $1 \leqslant p < \infty$, Γ *is a submanifold of an* $(n-1)$*-dimensional manifold* K *of smoothness* $k \geqslant 2$, $\bar\Gamma$ *is compact,* $\bar\Gamma \subset K$, $\eta > 0$, $\epsilon > 0$ *and there is given on the domain* $K^{(n)}_{+\eta} = K^{(n)}_{+\eta}(K)$ *a function* f *together with its weak derivatives up to order* λ *inclusively,* $1 \leqslant \lambda \leqslant k-1$, *such that for some* $\beta \geqslant 0$

$$\|f^{(\lambda)}_{x_1^{\lambda_1}\ldots x_n^{\lambda_n}}\|^{(n)}_{p(r,\,\beta)\,K^{(n)}_{+\eta}}<\infty \quad \text{for all} \quad \lambda_1+\cdots+\lambda_n=\lambda.$$

Suppose, further, $\beta = \tilde\beta p + \alpha$, $\tilde\beta$ *is a nonnegative integer,* $0 \leqslant \alpha < p$ *and* $\tilde\beta \leqslant \lambda - 1$. *Then for any* $\eta' \in (0, \eta)$ *there exists a function* F *defined on all of* E^n *and having the following properties.*

1°. $F \equiv f$ *on* $\Gamma^{(n)}_{+\eta'} = K^{(n)}_{+\eta'}(\Gamma)$.

2°. $F \in H^{(r)}_{p(n)}(M, E^n)$, $\|F\|^{(n)}_p \leqslant M$, $r = \lambda - \tilde\beta - (\alpha + \epsilon)/p$, (12.1)

$$M \leqslant c_1 \sum_{l=1}^{\lambda}\sum_{l_1+\cdots+l_n=\lambda-l}\left\|\frac{\partial f^{(\lambda-l)}}{\partial x_1^{l_1}\ldots\partial x_n^{l_n}}\right\|^{(n)}_{p\,K^{(n-1)}_{+\eta}}+c_2\sum_{\lambda_1+\cdots+\lambda_n=\lambda}\left\|\frac{\partial^\lambda f}{\partial x_1^{\lambda_1}\ldots\partial x_n^{\lambda_n}}\right\|^{(n)}_{p(r,\beta)\,K^{(n)}_{+\eta}}. \tag{12.2}$$

3°. *For any* $\xi \in [0, \eta')$

$$\left.\frac{\partial^\mu F}{\partial x_1^{\mu_1}\ldots\partial x_n^{\mu_n}}\right|_{\Gamma^{(n-1)}_{+\xi}}=\left.\frac{\partial^\mu f}{\partial x_1^{\mu_1}\ldots\partial x_n^{\mu_n}}\right|_{\Gamma^{(n-1)}_{+\xi}}\in H^{\left(r-\mu-\frac{1}{p}\right)}_{p\,(n-1)}(M_\xi,\Gamma^{(n-1)}_{+\xi}),$$

$$\Gamma^{(n-1)}_{+\xi}=K^{(n-1)}_{+\xi}(\Gamma),\ \mu=0,1,\ldots,\lambda-\tilde\beta-1, \qquad \textit{provided} \quad r-\mu-\frac{1}{p}>0.$$

Here M_ξ *satisfies an inequality of type* (12.2) *but generally with other constants also not depending on* f.

If $\alpha \neq p-1$ *and* $\tilde\beta = \lambda - 1$, *it is possible to take* $\epsilon = 0$.

A sufficient condition for condition (12.3) *is the condition*

$$1-\frac{\alpha+1+\varepsilon}{p}>0. \tag{12.4}$$

COROLLARY. *Suppose G is a finite domain in E^n bounded by an $(n-1)$-dimensional manifold K of smoothness $k \geqslant 2$ (or by a finite number of such manifolds) and suppose there is given in G (i) a continuous nonnegative function $\sigma = \sigma(P)$, there existing constants $\eta > 0$, $a_1 > 0$ and $a_2 > 0$ for which*

$$a_1 r(P) \leqslant \sigma(P) \leqslant a_2 r(P), \quad P \in K_\eta^{(n)}(K),\text{[6]}$$

and (ii) a function f together with its weak derivatives up to order λ inclusively, $1 \leqslant \lambda \leqslant k-1$, such that for some $\beta > 0$

$$\| f^{(\lambda)}_{x_1^{\lambda_1}\ldots x_n^{\lambda_n}} \|^{(n)}_{\substack{p(\sigma,\beta)\\ G}} < \infty \quad \text{for all} \quad \lambda_1 + \ldots + \lambda_n = \lambda.$$

Suppose, further, $\beta = \tilde{\beta} p + \alpha$, $0 \leqslant \alpha < p$, $\tilde{\beta}$ is a nonnegative integer, $\tilde{\beta} \leqslant \lambda - 1$, $r = \lambda - \tilde{\beta} - p^{-1}(\alpha + \epsilon)$ and

$$1 - \frac{\alpha + 1 + \varepsilon}{p} > 0.$$

Then f takes a value $f|_K$ on the boundary in the sense of convergence in the mean such that

$$\frac{\partial^\mu f}{\partial x_1^{\mu_1}\ldots \partial x_n^{\mu_n}}\bigg|_K \in H_{p\,(n-1)}^{\left(r-\mu-\frac{1}{p}\right)}(K), \quad \mu = 0, 1, \ldots, \lambda - \tilde{\beta} - 1.$$

If $\alpha \neq p-1$ and $\tilde{\beta} = \lambda - 1$, it is possible to take $\epsilon = 0$.

EXAMPLE. Suppose $\lambda = 1$, $0 \leqslant \alpha < 1$ and $p = 2$. Then $f|_K$ exists and satisfies the condition $f|_K \in H^{(\rho)}_{2(n-1)}$, where $\rho = 1 - \alpha/2 - 1/2 > 0$.

PROOF. Let $\{V^{(\nu)}\}$, $\nu = 1, \cdots, \nu_0$, be a finite covering of the compact set $\bar{\Gamma}$ by canonical neighborhoods of height η such that $V^{(\nu)} \subset K_\eta^{(n)} = K_\eta^{(n)}(K)$, and let $x_{(\nu)}(u)$ be the corresponding canonical mappings. Then the functions $f(x_{(\nu)}(u))$ are defined on $\overset{+}{V}{}_u^{(\nu)}$, and

$$\left\| \frac{\partial^\lambda f(x_{(\nu)}(u))}{\partial u_1^{\lambda_1}\ldots \partial u_n^{\lambda_n}} \right\|^{(n)}_{\substack{p(u_n\,\beta)\\ \overset{+}{V}{}_u^{(\nu)}}} < \infty \quad \text{for all} \quad \lambda_1 + \ldots + \lambda_n = \lambda.$$

---

(6) Here $r(P)$ is the distance along the normal from P to the manifold K.

Hence according to Theorems [11.1] and [11.2]

$$\left\| \frac{\partial^{\lambda-\widetilde{\beta}} f(x_{(\nu)}(u))}{\partial u_1^{\lambda_1'} \ldots \partial u_n^{\lambda_n'}} \right\|^{(n)}_{p\left(u_n,\, \alpha+\frac{\varepsilon}{2}\right)\, \overset{+}{V}{}^{(\nu)}_u} \leqslant M_\nu, \left\| \frac{\partial^{\lambda-\widetilde{\beta}-1} f(x_{(\nu)}(u))}{\partial u_1^{\kappa_1} \ldots \partial u_n^{\kappa_n}} \right\|^{(n)}_{p\, \overset{+}{V}{}^{(\nu)}_u} \leqslant M_\nu \tag{12.5}$$

for all $\lambda_1' + \cdots + \lambda_n' = \lambda - \beta$ and $\kappa_1 + \cdots + \kappa_n = \lambda - \widetilde{\beta} - 1$, where

$$M_\nu \leqslant c_\nu' \sum_{l=1}^{\widetilde{\beta}+1} \sum_{l_1+\ldots+l_n=\lambda-l} \left\| \frac{\partial^{\lambda-l} f(x_{(\nu)}(u))}{\partial u_1^{l_1} \ldots \partial u_n^{l_n}} \right\|^{(n-1)}_{p\, K^{(n-1)}_{+\eta}\left(\overset{\bullet}{V}{}^{(\nu)}_u\right)} + c_\nu'' \sum_{\lambda_1+\ldots+\lambda_n=\lambda} \left\| \frac{\partial^{\lambda} f(x_{(\nu)}(u))}{\partial u_1^{\lambda_1} \ldots \partial u_n^{\lambda_n}} \right\|^{(n)}_{p(u_n,\beta)\, \overset{+}{V}{}^{(\nu)}_u}; \tag{12.6}$$

All of the norms of the first sum are finite by virtue of Theorem [11.6] and the Remark to it, while all of the norms of the second sum are finite by assumption. Moreover, all of the derivatives of f of orders $\rho = 0, 1, \cdots, \lambda - \widetilde{\beta} - 1$ are p power summable on $\overset{+}{V}{}^{(\nu)}_u$.

In fact, from (12.5), by successively applying the Remark to Theorem [11.6] and Theorem [11.2] (for $\alpha = 0$) we get

$$\begin{aligned} \left\| \frac{\partial^{\lambda-\widetilde{\beta}-1-s} f(x_{(\nu)}(u))}{\partial u_1^{s_1} \ldots \partial u_n^{s_n}} \right\|^{(n)}_{p\, \overset{+}{V}{}^{(\nu)}_u} &\leqslant a_\nu' \sum_{t=1}^{s} \sum_{t_1+\ldots+t_n=\lambda-\widetilde{\beta}-1-t} \left\| \frac{\partial^{\lambda-\widetilde{\beta}-1-t} f(x_{(\nu)}(u))}{\partial u_1^{t_1} \ldots \partial u_n^{t_n}} \right\|^{(n-1)}_{p\, K^{(n-1)}_{+\eta}\left(\overset{\bullet}{V}{}^{(\nu)}_u\right)} \\ &+ a_\nu'' \sum_{\kappa_1+\ldots+\kappa_n=\lambda-\widetilde{\beta}-1} \left\| \frac{\partial^{\lambda-\widetilde{\beta}-1} f(x_{(\nu)}(u))}{\partial u_1^{\kappa_1} \ldots \partial u_n^{\kappa_n}} \right\|^{(n)}_{p\, \overset{+}{V}{}^{(\nu)}_u} < \infty. \end{aligned} \tag{12.7}$$

Thus the function $f(x_{(\nu)}(u))$ satisfies all of the conditions of Theorem [11.14] for $\bar{r} = \lambda - \widetilde{\beta} - 1$ and can therefore be extended onto all of the domain $V^{(\nu)}_u$ as a function $F_{(\nu)}$ satisfying (i) the condition

$$F_{(\nu)} \in H^{(r)}_{p\,(n)}(M_\nu', V^{(\nu)}_u),\ r = \bar{r} + 1 - \frac{\alpha+\varepsilon}{p} = \lambda - \widetilde{\beta} - \frac{\alpha+\varepsilon}{p}, \tag{12.8}$$

where M_ν' has the form of (12.6) under an appropriate choice of the constants c_ν' and c_ν'', and (ii) by virtue of (12.7) and formulas (6.5) the condition

$$\|F_{(\nu)}\|^{(n)}_{p\,V^{(\nu)}_u} \leqslant b'_\nu \sum_{t=1}^{\lambda-\widetilde{\beta}-1} \sum_{t_1+\cdots+t_n=\lambda-\widetilde{\beta}-1-t} \left\| \frac{\partial^{\lambda-\widetilde{\beta}-1-t} f(x_{(\nu)}(u))}{\partial u_1^{t_1}\cdots \partial u_n^{t_n}} \right\|^{(n-1)}_{p\,K^{n-1}_{+\eta}(\overset{\circ}{V}{}^{\nu}_u)} + b''_\nu \sum_{\kappa_1+\cdots+\kappa_n=\lambda-\widetilde{\beta}-1} \left\| \frac{\partial^{\lambda-\widetilde{\beta}-1} f(x_{(\nu)}(u))}{\partial u_1^{\kappa_1}\cdots \partial u_n^{\kappa_n}} \right\|^{(n)}_{p\,\overset{+(\nu)}{V_u}}. \tag{12.9}$$

Let $F^*_{(\nu)}(P) = F_{(\nu)}(x_{(\nu)}(u))$ for $P = x_{(\nu)}(u) \in V^{(\nu)}$. Clearly, $f(P) = F^*_{(\nu)}(P)$ for $P \in V^{(\nu)}$. We choose a finite system of canonical neighborhoods $\{U^{(\mu)}\}$, $\mu = 1, \cdots, \mu_0$, of height $\eta'' \in (\eta', \eta)$, forming a covering of $\overline{\Gamma}$ and such that for each μ there exists a $\nu = \nu(\mu)$ for which $\overline{U}^{(\mu)} \subset V^{(\nu)}$. Let $F^*_{(\mu)}$ denote the function defined on $U^{(\mu)}$ and coinciding with $F^*_{(\nu(\mu))}$. Then according to the theorem of Nikol'skiĭ on the behavior of the H-classes under a change of variables ([4], page 59) the conditions (12.8), (12.6), (12.9) and (12.5) imply

$$F^*_{(\mu)} \in H^{(r)}_{p\,(n)}(M^*, U^{(\mu)}), \quad \|F^*_{(\mu)}\|^{(n)}_{p\,U^{(\mu)}} \leqslant M^*, \quad \mu = 1, 2, \ldots, \mu_0, \tag{12.10}$$

where

$$M^* \leqslant d_1 \sum_{l=1}^{\lambda} \sum_{l_1+\cdots+l_n=\lambda-l} \left\| \frac{\partial^{\lambda-l} f}{\partial x_1^{\lambda_1}\cdots\partial x_n^{\lambda_n}} \right\|^{(n-1)}_{p\,K^{(n-1)}_{+\eta}} + d_2 \sum_{\lambda_1+\cdots+\lambda_n=\lambda} \left\| \frac{\partial^{\lambda} f}{\partial x_1^{\lambda_1}\cdots\partial x_n^{\lambda_n}} \right\|^{(n)}_{p(r,\beta)\,K^{(n)}_{+\eta}} \tag{12.11}$$

(for the sake of simplicity we have combined both constants and, by enlarging the domain of integration, have gotten rid of the index μ). Finally, we choose another system $\{W^{(\omega)}\}$, $\omega = 1, \cdots, \omega_0$, of canonical neighborhoods of height $\eta''' \in (\eta', \eta'')$, forming a covering of $\overline{\Gamma}$ and such that for each ω there exists a $\mu = \mu(\omega)$ for which $\overline{W}^{(\omega)} \subset U^{(\mu)}$. Let

$$W = \bigcup_{\omega=1}^{\omega_0} W^{(\omega)}, \quad V = \bigcup_{\nu=1}^{\nu_0} V^{(\nu)}, \quad U = \bigcup_{\mu=1}^{\mu_0} U^{(\mu)}.$$

Then $\overline{\Gamma} \subset W \subset \overline{W} \subset U \subset \overline{U} \subset V$. Now, by applying the usual method, we can smooth out the functions $F^*_{(\mu)}$ into a single function F coinciding with f on W, equal to zero outside V and satisfying conditions 1° and 2° of the theorem. This is done by the same construction that was described in Theorem [12.2]. In this connection (12.1) follows from (12.8) while (12.2) follows from (12.11). Finally, assertion 3° of the theorem follows from 1° and 2° according to an imbedding theorem of Nikol'skiĭ ([4], page 104).

[12.4]. *Suppose* Γ *is a submanifold of an* $(n-1)$*-dimensional manifold* K *of smoothness* $k \geqslant 2$, $\overline{\Gamma}$ *is compact,* $\overline{\Gamma} \subset K$, $0 \leqslant \alpha < p$ *and on the domain* $K^{(n)}_{+\eta}(\Gamma)$ *there is given a function* f *together with all of its weak derivatives of first order such that*

$$\sum_{i=1}^{n} \| f_{x_i} \|^{(n)}_{p(r,\alpha),\, K^{(n)}_{+\eta}(\Gamma)} < \infty.$$

Then there exists a constant $A > 0$ *such that*

$$\| f \|^{(n)}_{p,\, K^{(n)}_{+\eta}(\Gamma)} \leqslant A \left[\| f \|^{(n-1)}_{p,\, K^{(n-1)}_{+\eta}(\Gamma)} + \sum_{i=1}^{n} \| f_{x_i} \|^{(n)}_{p(r,\alpha),\, K^{(n)}_{+\eta}(\Gamma)} \right].$$

[12.5]. *Suppose* Γ *is a submanifold of an* $(n-1)$*-dimensional manifold* K *of smoothness* $k \geqslant 2$, $\overline{\Gamma}$ *is compact,* $\overline{\Gamma} \supset K$, $0 < \alpha < p-1$ *and on the domain* $K^{(n)}_{+\eta}(K)$ *there is given a function* f *together with all of its weak derivatives of first order such that*

$$\sum_{i=1}^{n} \| f_{x_i} \|^{(n)}_{p(r,\alpha),\, K^{(n)}_{+\eta}(\Gamma)} < \infty, \quad f|_{\Gamma} = 0.$$

Then there exists a constant $B > 0$ *not depending on* f *such that for* $0 < h < \eta$

$$\| f \|^{(n)}_{p(r,\alpha),\, K^{(n)}_{+h}(\Gamma)} \leqslant hB \sum_{i=1}^{n} \| f_{x_i} \|^{(n)}_{p(r,\alpha),\, K^{(n)}_{+\eta}(\Gamma)}.$$

[12.6]. *Suppose* Γ *is a submanifold of an* $(n-1)$*-dimensional manifold* K *of smoothness* $k \geqslant 2$, $\overline{\Gamma}$ *is compact,* $\overline{\Gamma} \subset K$, $0 \leqslant \alpha < p-1$ *and on the domain* $K^{(n)}_{+\eta}(\Gamma)$ *there is given a function* f *together with all of its weak derivatives of first order such that*

$$\sum_{i=1}^{n} \| f_{x_i} \|^{(n)}_{\substack{p\,(r,\,\alpha) \\ K^{(n)}_{+\eta}(\Gamma)}} < \infty .$$

Then there exists a constant $c > 0$ *not depending on* f *such that*

$$\left| \| f \|^{(n-1)}_{\substack{p \\ K^{(n-1)}_{+\eta}(\Gamma)}} - \| f \|^{(n-1)}_{\substack{p \\ \Gamma}} \right| \leqslant c \sum_{i=1}^{n} \| f_{x_i} \|^{(n)}_{\substack{p\,(r,\,\alpha) \\ K^{(n)}_{+\eta}(\Gamma)}} .$$

The proof of the last three assertions follows directly from Theorems [11.2], [11.10] and [11.11] if one goes over to canonical neighborhoods by means of canonical mappings. The finiteness of all of the considered norms follows from Theorem [12.3].

[12.7]. DEFINITION. Suppose G is a domain in E^n, K is its boundary and Γ is a finite number of pairwise disjoint $(n-1)$-dimensional manifolds such that $\Gamma \subset K$. We will say that the domain G is *regular relative to* Γ if there exists a partitioning of G into a finite number of domains $G^{(\nu)}$. with boundaries of zero measure and a partitioning of Γ into a finite number of manifolds $\Gamma^{(\nu)}$, $\nu = 1, \cdots, \nu_0$: $\bar{G} = \bigcup_1^{\nu_0} \bar{G}^{(\nu)}$, $\bar{\Gamma} = \bigcup_1^{\nu_0} \bar{\Gamma}^{(\nu)}$, having the property that for each ν there exists a number $i_\nu = 1, \cdots, n$ such that the intersection with $G^{(\nu)}$ of every straight line parallel to the Ox_{i_ν} axis and passing through a point of $G^{(\nu)}$ is an interval one end of which lies on $\Gamma^{(\nu)}$, thus implying that $\Gamma^{(\nu)}$ has an explicit representation relative to the x_{i_ν} coordinate.

We will also say that the domain G is *regular relative to* Γ if a similar partitioning is possible relative to a curvilinear coordinate system. Clearly, this requirement is not in a sense essentially restrictive for the class of domains with a piecewise smooth boundary.

A finite domain G bounded by an $(n-1)$-dimensional manifold K (or by a finite number of such manifolds) will be called a *simple domain* if there exists an $\eta > 0$ such that the domain G_η whose boundary K_η is obtained by an η displacement of K along the normals toward the interior of G, i.e. satisfies the condition $K_\eta = K^{(n-1)}_{+\eta}(K)$, is regular relative to K_η.

In the sequel, without specifically mentioning it, we will assume that every domain G bounded by a finite number of pairwise disjoint $(n-1)$-dimensional manifolds is simple.

We note, for example, that the plane domain whose boundary consists of the graph Γ_1 of the function $y = \sin x^{-1}$ for $0 < x < \pi^{-1}$, the segment Γ_2 with endpoints $(0, -1)$ and $(0, 2)$, the interval Γ_3 with endpoints $(0, 2)$ and $(\pi^{-1}, 2)$ and the segment Γ_4 with endpoints $(\pi^{-1}, 2)$ and $(\pi^{-1}, 0)$ is regular relative to Γ_3, although it is not a Ljapunov domain.

For a function $f(x, y)$ defined on this domain it is possible to prove completely analogously to [11.2] the inequality

$$\|f\|_{\substack{p\\G}}^{(2)} \leqslant c_1 \|f\|_{\substack{p\\\Gamma_3}}^{(1)} + c_2 \|f_y\|_{\substack{p(r,\alpha)\\G}}^{(2)}, \quad r = 2 - y, \quad 0 \leqslant \alpha < p - 1.$$

[12.8]. *Suppose G is a finite domain bounded by an $(n-1)$-dimensional manifold K of smoothness $k \geqslant 2$, $0 \leqslant \alpha < p - 1$ and $\sigma = \sigma(P)$ is a continuous positive function on G such that there exist constants $\eta > 0$, $a_1 > 0$ and $a_2 > 0$ for which*

$$a_1 r(P) \leqslant \sigma(P) \leqslant a_2 r(P), \quad P \in K_{+\eta}^{(n)}(K). \tag{12.12}$$

Suppose, further, there is defined on G a function f together with all of its weak derivatives of first order such that $\sum_{i=1}^n \|f_{x_i}\|_{\substack{p(\sigma,\alpha)\\G}}^{(n)} < \infty$. Then there exist constants $c_1 > 0$ and $c_2 > 0$, not depending on f, such that

$$\|f\|_{\substack{p\\G}}^{(n)} \leqslant c_1 \|f\|_{\substack{p\\K}}^{(n-1)} + c_2 \sum_{i=1}^{n} \|f_{x_i}\|_{\substack{p(\sigma,\alpha)\\G}}^{(n)}.$$

PROOF. From the conditions of the theorem and the Corollary to [12.3] (for $\lambda = 1$) it follows that $\|f\|_{\substack{p\\K}}^{(n-1)} < \infty$. We choose an $\eta > 0$ and a partioning of the domain G_η into subdomains $G^{(\nu)}$ according to definition [12.7]. Then, putting $K_{+\eta}^{(n)}(K) = K_{+\eta}^{(n)}$ and $K_{+\eta}^{(n-1)}(K) = K_{+\eta}^{(n-1)}$, we will have

$$\|f\|_{\substack{p\\G}}^{(n)} \leqslant \|f\|_{\substack{p\\K_{+\eta}^{(n)}}}^{(n)} + \sum_{\nu=1}^{\nu_0} \|f\|_{\substack{p\\G^{(\nu)}}}^{(n)}.$$

From Theorems [12.4] and [12.6] we get

$$\|f\|^{(n)}_{p\,K^{(n)}_{+\eta}} \leqslant A\Big[\|f\|^{(n-1)}_{p\,K^{(n-1)}_{+\eta}} + \sum_{i=1}^{n}\|f_{x_i}\|^{(n)}_{p(r,\alpha)\,K^{(n)}_{+\eta}}\Big]$$

$$\leqslant A\|f\|^{(n-1)}_{p\,K} + A(c+1)\sum_{i=1}^{n}\|f_{x_i}\|^{(n)}_{p(r,\alpha)\,K^{(n)}_{+\eta}}$$

$$\leqslant b_1\|f\|^{(n-1)}_{p\,K} + b_2\sum_{i=1}^{n}\|f_{x_i}\|^{(n)}_{p(\sigma,\alpha)\,G}.$$

We next take an arbitrary ν and suppose for the sake of definiteness that $i_\nu = n$ (see [2.7]). We denote by $G^{(\nu)}_{x_n}$ the orthogonal projection of $G^{(\nu)}$ onto the hyperplane $x_n = 0$ and for each point $P = (x_1, \cdots, x_{n-1}) \in G^{(\nu)}_{x_n}$ we denote by $\varphi(x_1, \cdots, x_{n-1})$ and $\psi(x_1, \cdots, x_{n-1})$ the ends of the interval of intersection of the straight line parallel to the Ox_n axis and passing through P with $G^{(\nu)}$ respectively belonging and not belonging to $\Gamma^{(\nu)}$. We will assume for the sake of definiteness that $\varphi \leqslant \psi$. By virtue of the finiteness of G there exists a constant $d > 0$ such that $0 \leqslant \psi - \varphi \leqslant d$. We now have

$$\|f\|^{(n)}_{p\,G^{(\nu)}} = \Big\{\int\!\cdots\!\int_{G^{(\nu)}} |f(x_1, \ldots, x_n)|^p\, dx_1 \ldots dx_n\Big\}^{\frac{1}{p}}$$

$$= \Big\{\int\!\cdots\!\int_{G^{(\nu)}} \Big|f(x_1, \ldots, x_{n-1}, \varphi(x_1, \ldots, x_{n-1})) + \int_{\varphi(x_1, \ldots, x_{n-1})}^{x_n} f_{x_n}(x_1, \ldots, x_{n-1}, t)\, dt\Big|^p dx_1 \ldots dx_n\Big\}^{\frac{1}{p}}$$

$$\leqslant \Big\{\int\!\cdots\!\int_{G^{(\nu)}_{x_n}} \Big(\int_{\varphi}^{\psi} dx_n\Big)|f(x_1, \ldots, x_{n-1}, \varphi)|^p\, dx_1 \ldots dx_{n-1}\Big\}^{\frac{1}{p}}$$

$$+ \Big\{\int\!\cdots\!\int_{G^{(\nu)}} \Big|\int_{\varphi}^{x_n} f_{x_n}(x_1, \ldots, x_{n-1}, t)\Big|^p dx_1 \ldots dx_n\Big\}^{\frac{1}{p}}$$

$$\leqslant d^{\frac{1}{p}}\Big\{\int\!\cdots\!\int_{G^{(\nu)}_{x_n}} |f(x_1, \ldots, x_{n-1}, \varphi)|^p\, dx_1 \ldots dx_{n-1}\Big\}^{\frac{1}{p}} +$$

$$+\Big\{\int_{G^{(\nu)}}\!\!\cdots\int\int_{\varphi}^{x_n}|f_{x_n}(x_1,\ldots,x_{n-1},t)|^p\,dt\,(x_n-\varphi)^{\frac{p}{q}}\,dx_1\ldots dx_n\Big\}^{\frac{1}{p}}$$

$$\leqslant d^{\frac{1}{p}}\Big\{\int_{\Gamma^{(\nu)}}\!\!\cdots\int|f(x_1,\ldots,x_{n-1},\varphi)|^p\,d\Gamma^{(\nu)}\Big\}^{\frac{1}{p}}$$

$$+\,d^{\frac{1}{q}}\Big\{\int_{G^{(\nu)}_{x_n}}\!\!\cdots\int\int_{\varphi}^{\psi}|f_{x_n}(x_1,\ldots,x_{n-1},t)|^p\Big(\int_{\varphi}^{\psi}dx_n\Big)\,dx_1,\ldots,dx_{n-1}dt\Big\}^{\frac{1}{p}} \tag{12.13}$$

$$\leqslant d^{\frac{1}{p}}\,\|f\|^{(n-1)}_{\substack{p\\ \Gamma^{(\nu)}}}+d\,\|f_{x_n}\|^{(n)}_{\substack{p\\ G^{(\nu)}}}.$$

Again applying [12.6], we get

$$\|f\|^{(n-1)}_{\substack{p\\ \Gamma^{(\nu)}}}\leqslant\|f\|^{(n-1)}_{\substack{p\\ K}}+c\sum_{i=1}^{n}\|f_{x_i}\|^{(n)}_{\substack{p\,(r,\,\alpha)\\ K^{(n)}_{+\eta}}}$$

$$\leqslant\|f\|^{(n-1)}_{\substack{p\\ K}}+c'\sum_{i=1}^{n}\|f_{x_i}\|^{(n)}_{\substack{p\,(\sigma,\,\alpha)\\ G}}.$$

Finally, if $\sigma_0=\min_{P\in G_\eta}\sigma(P)$ (clearly, $\sigma_0>0$), then

$$\|f_{x_n}\|^{(n)}_{\substack{p\\ G^{(\nu)}}}\leqslant\Big\{\int_{G^{(\nu)}}\!\!\cdots\int\Big(\frac{\sigma}{\sigma_0}\Big)^{\alpha}\Big|f_{x_n}(x_n,\ldots,x_n)\Big|^p dx_1\ldots dx_n\Big\}^{\frac{1}{p}}$$

$$\leqslant\sigma_0^{-\frac{\alpha}{p}}\|f_{x_n}\|^{(n)}_{\substack{p\,(\sigma,\,\alpha)\\ G^{(\nu)}}}\leqslant\sigma_0^{-\frac{\alpha}{p}}\sum_{i=1}^{n}\|f_{x_i}\|^{(n)}_{\substack{p\,(\sigma,\,\alpha)\\ G}}.$$

The above inequalities prove Theorem [12.8].

This theorem carries over to the case of an infinite layer.

[12.9]. *Let*

$$\overset{+}{E}{}^n_a=\{(x_i):0<x_n<a\},\ E^{n-1}=\{(x_i):x_n=0\},\ 0\leqslant\alpha<p-1.$$

and suppose there is defined on $\overset{+}{E}{}^n_a$ *a function* f *together with all of its weak derivatives of first order such that*

$$\sum_{i=1}^{n} \|f_{x_i}\|^{(n)}_{\substack{p(x_n,\alpha)\\ E^{+n}_a}} < \infty, \quad \|f\|^{(n-1)}_{\substack{p\\ E^{n-1}}} < \infty.$$

Then there exist constants $c_1 > 0$ *and* $c_2 > 0$ *not depending on* f *such that*

$$\|f\|^{(n)}_{\substack{p\\ G}} \leqslant c_1 \|f\|^{(n-1)}_{\substack{p\\ E^{n-1}}} + c_2 \sum_{i=1}^{n} \|f_{x_i}\|^{(n)}_{\substack{p(x_n,\alpha)\\ E^{+n}_a}}.$$

The proof is analogous to the proof of Theorem [12.8].

§13. Completeness of weighted spaces

Let Γ be a finite domain in the hyperplane $E^{n-1} = \{(x_i): x_n = 0\}$ that possibly coincides with E^{n-1}, and let

$$G = \{(x_i): (x_1, \ldots, x_{n-1}) \in \Gamma, \ 0 < x_n < a \leqslant \infty\}.$$

Further, let $\alpha_\lambda \geqslant 0$, $\lambda = 0, 1, \cdots, r$, and suppose there is defined on G a function f together with all of its weak derivatives up to order r inclusively such that

$$\left\| \frac{\partial^\lambda f}{\partial x_1^{\lambda_1} \ldots \partial x_n^{\lambda_n}} \right\|^{(n)}_{p(x_n,\alpha_\lambda)} < \infty$$

for all $\lambda_1 + \cdots + \lambda_n = \lambda$, $\lambda = 0, 1, \cdots, r$. We put

$$\|f\|^{(n)}_{W^{(r)}_p(x_n;\,\alpha_0,\ldots,\alpha_r;\,G)} = \sum_{\lambda=0}^{r} \sum_{\lambda_1+\ldots+\lambda_n=\lambda} \left\| \frac{\partial^\lambda f}{\partial x_1^{\lambda_1} \ldots \partial x_n^{\lambda_n}} \right\|^{(n)}_{\substack{p(x_n,\alpha_\lambda).\\ G}} \tag{13.1}$$

The defined norm satisfies all of the necessary conditions:

$$\|\mu_1 f_1 + \mu_2 f_2\| \leqslant |\mu_1| \|f_1\| + |\mu_2| \|f_2\|$$

as follows from the weighted Minkowski inequality ([2], page 6), and $f = 0$ almost everywhere on G if $\|f\| = 0$.

Thus the set of all functions f for which the norm (13.1) is finite forms a normed linear space which we will denote by $W^{(r)}_p(x_n; \alpha_0, \cdots, \alpha_r; G)$ or simply $W^{(r)}_p(x_n; \alpha_0, \cdots, \alpha_r)$. This space is a generalization of the space $W^{(r)}_p$ introduced and studied by Sobolev [2] (see [13.2] below).

[13.1]. *The normal linear space* $W_p^{(r)}(x_n; \alpha_1, \cdots, \alpha_r; G)$ *is complete.*

For if a sequence of functions $f_k \in W_p^{(r)}(x_n; \alpha_0, \cdots, \alpha_r)$ forms a Cauchy sequence in the sense of the norm (13.1), it can be established in exactly the same way as in the case of the ordinary space $L_p^{(n)}$ that for every considered system of numbers $\lambda_1, \cdots, \lambda_n$ $(\lambda_1 + \cdots + \lambda_n = \lambda = 0, 1, \cdots, r)$ there exists a function $\varphi_{\lambda_1,\dots,\lambda_n}$ such that

$$\|\varphi_{\lambda_1 \dots \lambda_n}\|^{(n)}_{p(x_n, \alpha_\lambda)} < \infty, \tag{13.2}$$
$$G$$

$$\lim_{k\to\infty} \left\| \frac{\partial^\lambda f_k}{\partial x_1^{\lambda_1} \dots \partial x_n^{\lambda_n}} - \varphi_{\lambda_1,\dots,\lambda_n} \right\|^{(n)}_{p(x_n, \alpha_\lambda)} = 0, \tag{13.3}$$

and hence for any subdomain D such that $\bar{D} \subset G$

$$\lim_{k\to\infty} \left\| \frac{\partial^\lambda f_k}{\partial x_1^{\lambda_1} \dots \partial x_n^{\lambda_n}} - \varphi_{\lambda_1,\dots,\lambda_n} \right\|^{(n)}_{p} = 0.$$
$$D$$

It follows according to a result of S. M. Nikol'skiĭ [4], Lemma 1.2, pages 43–44) that

$$\varphi_{\lambda_1,\dots,\lambda_n} = \frac{\partial^\lambda \varphi_{0\dots0}}{\partial x_1^{\lambda_1}, \dots, \partial x_n^{\lambda_n}} \quad \text{for all} \quad \lambda_1 + \dots + \lambda_n = \lambda. \tag{13.4}$$

Relations (13.2) and (13.4) imply that $\varphi_{0\dots0} \in W_p^{(r)}(x_n: \alpha_0, \cdots, \alpha_r; G)$, while relation (13.3) implies that $\lim_{k\to\infty} f_k = \varphi_{0\dots0}$ in the sense of the norm (13.1).

The space $W_p^{(r)}(r; \alpha_0, \cdots, \alpha_r; G)$ in the case when the domain Γ lies in an m-dimensional hyperplane E^m, $1 \leqslant m < n-1$, and $G = \Gamma_a^{(n)}$ (see §11) can be considered in exactly the same way.

REMARK 1°. In the case $p = 2$ the space $W_p^{(r)}(x_n; \alpha_0, \cdots, \alpha_r)$ can be converted into a Hilbert space by introducing the scalar product

$$(f, g) = \sum_{\lambda=0} \sum_{\lambda_1+\dots+\lambda_n=\lambda} \int_G \dots \int x_n^{\alpha_\lambda} f^{(\lambda)}_{x_1^{\lambda_1} \dots x_n^{\lambda_n}}(x_1, \dots, x_n) \times g^{(\lambda)}_{x_1^{\lambda_1},\dots, x_n^{\lambda_n}}(x_1, \dots, x_n)\, dx_1 \dots dx_n. \tag{13.5}$$

[13.2]. *Suppose* G *is a bounded domain,* $r \geqslant 1, \alpha_r = \alpha = (r-1) + \beta, 0 \leqslant \beta < p$, $\epsilon > 0, \beta + \epsilon(r-1) < p, \alpha_\lambda = \alpha - (r-\lambda)p + (r-\lambda)\epsilon, \lambda = 1, 2, \cdots, r, \alpha_0 = 0$, *and* a *is finite. If*

$$\| f^{(r)}_{x_1^{r_1}, \ldots, x_n^{r_n}} \|^{(n)}_{\substack{p(x_n, \alpha) \\ G}} < \infty$$

for all $r_1 + \cdots + r_n = r$, *then* $f \in W_p^{(r)}(x_n; \alpha_0, \cdots, \alpha_r)$.

This immediately follows from [11.1], [11.2] and the Remark to [11.6].

Thus in order to establish the membership of f in $W_p^{(r)}(x_n; \alpha_0, \cdots, \alpha_r)$ under the indicated choice of exponents $\alpha_0, \cdots, \alpha_r$ it suffices to know the behavior of the leading derivatives of f. This choice of the exponents $\alpha_0, \cdots, \alpha_{r-1}$ for a given α_r is therefore the most natural to within $\epsilon > 0$.

REMARK 2°. Suppose now G is a finite domain bounded by an $(n-1)$-dimensional manifold K of smoothness $k \geqslant 2, r \leqslant k-1$ and $\sigma = \sigma(p)$ is a continuous positive function on G satisfying condition (12.12). Then, analogously to the way it was done above, we can define a normed linear space $W_p^{(r)}(\sigma; \alpha_0, \cdots, \alpha_r; G)$ for which natural modifications of propositions [13.1] and [13.2] are valid.

§14. Limit theorems

Let $E^m = \{(x_i): x_{m+1} = \cdots = x_n = 0\}$. For the sake of brevity we put $\mathring{f} = f(x_1, \cdots, x_m, 0, \cdots, 0)$ (in the sense of the mean value) for every function f defined on E^n.

[14.1]. *Suppose there is given in* E^n *a sequence of functions* $F_k \in H_p^{(r_1, \ldots, r_n)}(M)$, $\|F_k\|_p^{(n)} \leqslant M$, $k = 1, 2, \cdots$, *and* $\chi = 1 - p^{-1}\sum_{m+1}^n 1/r_i > 0$. *Suppose further,* Γ *is a bounded domain in the hyperplane* E^m,

$$G = \{(x_i): (x_1, \ldots, x_m) \in \Gamma, |x_j| < a < \infty, j = m+1, \ldots, n\},$$

the sequence F_k *converges to a function* F *in* $L_p^{(n)}(G)$ *and the sequence* $\mathring{F}_k$ *converges to a function* f *in* $L_p^{(m)}(\Gamma)$. *Then* $\mathring{F}$ *exists and is equal to* f *on* Γ.(7)

PROOF. We employ convergent (in $L_p^{(n)}$) expansions of the F_k in entire functions: $F_k = \sum_{\nu=0}^\infty q_{\nu k}$, where the $q_{\nu k}$ are entire functions of exponential type of degree $2^{\nu/r_i}$ in $x_i, i = 1, \cdots, n$, such that

---

(7) This assertion is a refinement of an earlier result of the author ([15], Lemma 1°).

$$\| q_{\nu k} \|_p^{(n)} \leqslant \frac{cM}{2^\nu}, \tag{14.1}$$

where the constant c does not depend on the number ν (see (2.16) and (2.17)).

Let $Q_{\mu k} = \Sigma_0^\mu Q_{\nu k}$. Clearly, $Q_{\mu k}$ is an entire function of degree $2^{\mu/r_i}$ in x_i $(i = 1, \cdots, n)$ and

$$\| Q_{\mu k} \|_p^{(n)} \leqslant \sum_{\nu=0}^{\mu} \| q_{\nu k} \|_p^{(n)} \leqslant \sum_{\nu=0}^{\infty} \frac{cM}{2^\nu} = 2cM,$$

which implies by virtue of the inequality

$$\| g_{\nu_1, \ldots, \nu_n} \|_{p'}^{(n)} \leqslant 2^n \left(\prod_{i=1}^{n} \nu_i \right)^{\frac{1}{p} - \frac{1}{p'}} \| g_{\nu_1, \ldots, \nu_n} \|_p^{(n)},$$

of S. M. Nikol'skiĭ ([3], page 10), which is valid for any entire function of exponential type of degree ν_i in x_i $(i = 1, \cdots, n)$ and $1 \leqslant p < p' \leqslant \infty$, that

$$\| Q_{\mu k} \|_\infty^{(n)} \leqslant 2^n \left(\prod_{i=1}^{n} 2^{\frac{\mu}{r_i}} \right)^{\frac{1}{p}} \| Q_{\mu k} \|_p^{(n)} \leqslant 2^{n+1} cM \left(\prod_{i=1}^{n} 2^{\frac{\mu}{r_i}} \right)^{\frac{1}{p}},$$

i.e. for fixed $\mu = 1, 2, \cdots$ the sequence of entire functions $Q_{\mu k}$ is bounded in absolute value in E^n. In this case, as is well known, it is possible to extract from the sequence $Q_{\mu k}$ a subsequence that converges uniformly on the compact set $\overline{G}$. By applying, if necessary, the diagonal process, we can obtain a sequence of indices k_s such that all of the corresponding sequences of functions $Q_{\mu k_s}$, $s = 1, 2, \cdots$, will converge uniformly on $\overline{G}$ in such a way that

$$\| Q_{\mu k_s} - Q_{\mu k_{s-1}} \|_\infty^{(n)} < \frac{1}{2^s}, \quad s = 1, 2, \ldots. \tag{14.2}$$

Further, applying (14.1), we have

$$\| F_{k_s} - Q_{k_s k_s} \|_p^{(n)} = \left\| \sum_{\nu = k_s + 1}^{\infty} q_{\nu k_s} \right\|_p^{(n)} \leqslant \sum_{\nu = k_s + 1}^{\infty} \frac{cM}{2^\nu} = \frac{cM}{2^{k_s}}. \tag{14.3}$$

According to another inequality of S. M. Nikol'skiĭ for entire functions (see (4.7) as well as [3], page 11) inequality (14.1) also implies

$$\|q_{\nu k}\|_p^{(m)} \leqslant 2^n \left(\prod_{i=m+1}^{n} 2^{\frac{\nu}{r_i}} \right)^{\frac{1}{p}} \|q_{\nu k}\|_p^{(n)} \leqslant \frac{2^n cM}{2^{\varkappa\nu}} \tag{14.4}$$

for any fixed $x_{m+1}, \cdots, x_n$. Therefore, noting that $\mathring{F}_k = \Sigma_{\nu=0}^{\infty} q_{\nu k}$, we will have

$$\left\| \mathring{F}_{k_s} - \mathring{Q}_{k_s k_s} \right\|_p^{(m)} = \left\| \sum_{\nu=k_s+1}^{\infty} \mathring{q}_{\nu k_s} \right\|_p^{(m)} \leqslant \sum_{\nu=k_s+1}^{\infty} \frac{2^n cM}{2^{\varkappa k_s}}. \tag{14.5}$$

From the conditions of the theorem and inequalities (14.3) and (14.5) we respectively get

$$\|F - Q_{k_s k_s}\|_{p,G}^{(n)} \leqslant \|F - F_{k_s}\|_{p,G}^{(n)} + \|F_{k_s} - Q_{k_s k_s}\|_p^{(n)} \to 0 \quad \text{for} \quad s \to \infty,$$

$$\left\| f - \mathring{Q}_{k_s k_s} \right\|_{p,\Gamma}^{(m)} \leqslant \left\| f - \mathring{F}_{k_s} \right\|_{p,\Gamma}^{(m)} + \left\| \mathring{F}_{k_s} - \mathring{Q}_{k_s k_s} \right\|_p^{(m)} \to 0 \quad \text{for} \quad s \to \infty.$$

Consequently, in $L_p^{(n)}(G)$

$$F = Q_{k_1 k_1} + \sum_{s=2}^{\infty} (Q_{k_s k_s} - Q_{k_{s-1} k_{s-1}}), \tag{14.6}$$

while in $L_p^{(m)}(\Gamma)$

$$f = \mathring{Q}_{k_1 k_1} + \sum_{s=2}^{\infty} \left(\mathring{Q}_{k_s k_s} - \mathring{Q}_{k_{s-1} k_{s-1}} \right). \tag{14.7}$$

Let $G_{x_{m+1} \cdots x_n}$ denote the intersection of G with the m-dimensional hyperplane obtained by fixing the values of $x_{m+1}, \cdots, x_n$. By virtue of (14.2) and (14.4) we have

$$\| Q_{k_s k_s} - Q_{k_{s-1} k_{s-1}} \|_{p, G_{x_{m+1} \cdots x_n}}^{(m)}$$

$$\leqslant \| Q_{k_s k_s} - Q_{k_s k_{s-1}} \|_{p, G_{x_{m+1} \cdots x_n}}^{(m)} + \| Q_{k_s k_{s-1}} - Q_{k_{s-1} k_{s-1}} \|_{p, G_{x_{m+1} \cdots x_n}}^{(m)}$$

$$\leqslant \| Q_{k_s k_s} - Q_{k_s k_{s-1}} \|_{\infty}^{(n)} \operatorname{mes}^{\frac{1}{p}} \Gamma + \sum_{\nu=k_{s-1}+1}^{k_s} \| q_{\nu k_{s-1}} \|_p^{(m)}$$

$$\leqslant \frac{1}{2^s} \operatorname{mes}^{\frac{1}{p}} \Gamma + \sum_{\nu=k_{s-1}+1}^{\infty} \frac{2^n cM}{2^{\varkappa\nu}}$$

$$= \frac{1}{2^s} \operatorname{mes}^{\frac{1}{p}} \Gamma + \frac{2^n cM}{2^{\varkappa k_{s-1}}} \to 0, \text{ for } s \to \infty. \tag{14.8}$$

The convergence to zero here occurs uniformly relative to the admissible values of $x_{m+1}, \cdots, x_n$.

Now, using expansions (14.6) and (14.7), we get for any natural λ that

$$[\int_\Gamma \cdots \int | F(x_1, \ldots, x_n) - f(x_1, \ldots, x_m) |^p dx_1 \ldots dx_m]^{\frac{1}{p}}$$

$$= \Big\{ \int_\Gamma \cdots \int \Big| Q_{k_1k_1} + \sum_{=2}^{\infty} (Q_{k_sk_s} - Q_{k_{s-1}k_{s-1}}) - \mathring{Q}_{k_1k_1}$$

$$+ \sum_{s=2}^{\infty} \left(\mathring{Q}_{k_sk_s} - \mathring{Q}_{k_{s-1}k_{s-1}} \right) \Big|^p dx_1 \ldots dx_m \Big\}^{\frac{1}{p}}$$

$$\leqslant \left(\int_\Gamma \cdots \int \left| Q_{k_\lambda k_\lambda} - \mathring{Q}_{k_\lambda k_\lambda} \right|^p dx_1 \ldots dx_m \right)^{\frac{1}{p}}$$

$$+ \sum_{s=\lambda+1}^{\infty} \left\| Q_{k_sk_s} - Q_{k_{s-1}k_{s-1}} \right\|_{p}^{(m)} {}_{G_{x_{m+1}\ldots x_n}} + \sum_{s=\lambda+1}^{\infty} \left\| \mathring{Q}_{k_sk_s} - \mathring{Q}_{k_{s-1}k_{s-1}} \right\|_{p}^{(m)} {}_{G_{x_{m+1}\ldots x_n}}.$$

Finally, suppose an $\epsilon > 0$ is given. We choose λ so large that the sum of the last two terms in the right side of the preceding inequality becomes less than $\epsilon/2$ independently of $x_{m+1}, \cdots, x_n$; this can be done on the basis of (14.8). After this, by virtue of the uniform continuity of the functions $Q_{k_\lambda k_\lambda}$ on $\overline{G}$ we can choose a $\delta > 0$ such that the first term will also be less than $\epsilon/2$ when $\Sigma_1^m |x_i| < \delta$. This means that $\mathring{F} = f$.

The method of diagonal sums $Q_{k_sk_s}$ employed in the proof of this theorem can also be used to estimate the class of the limit theorem.

[14.2]. *Suppose given in* E^n *a sequence of functions* $F_s \in H_p^{(r_1, \cdots, r_n)}(M)$, $\|F_s\|_p^{(n)} \leqslant M$, $s = 0, 1, 2, \cdots$, *converging in* $L_p^{(n)}$ *to a function* F. *Then* $F \in H_p^{(r_1, \cdots, r_n)}(aM)$, *where the constant* $a > 0$ *does not depend on* M.

PROOF. Going over, if necessary, to a subsequence, we can assume with loss of generality that

$$\| F - F_s \|_p^{(n)} < \frac{M}{2^s}, \quad s = 0, 1, \ldots .$$

We again consider expansions of the F_s in entire functions $q_{\nu s}$ of degree $2^{\nu/r_i}$ in x_i $(i = 1, \cdots, n)$: $F_s = \Sigma_0^\infty q_{\nu s}$, so that conditions (14.1) are satisfied, and let, as above, $Q_s = \Sigma_0^s q_{\nu s}$, $s = 0, 1, \cdots$. Analogously to the way inequality (14.3) (for $k_s = s$) was obtained, we have

$$\| F_s - Q_s \|_p^{(n)} \leqslant \frac{cM}{2^s},$$

where the constant $c > 0$ does not depend on the functions F_s. Hence

$$\| F - Q_s \|_p^{(n)} \leqslant \| F - F_s \|_p^{(n)} + \| F_s - Q_s \|_p^{(n)} \leqslant \frac{M(c+1)}{2^s}. \tag{14.9}$$

The function Q_s is an entire function of degree $\nu_s = 2^{s/r_i}$ in x_i $(i = 1, \cdots, n; s = 0, 1, \cdots)$. Therefore, from (14.9), for the best approximations $A_{\nu_s x_i}(F)_p$ of F by entire functions of degree ν_s in x_i we have

$$A_{\nu_s x_i}(F)_p \leqslant \frac{M(c+1)}{\nu_s^{r_i}}, \quad i = 1, 2, \ldots, n.$$

In addition,

$$\| F_s \|_p^{(n)} \leqslant \| F \|_p^{(n)} + \| F - F_s \|_p^{(n)} \leqslant M + \frac{M}{2^s} \leqslant 2M.$$

But then by virtue of the corresponding theorem of Nikol'skiĭ ([3], Theorem 8, page 21) we will have

$$F \in H_p^{(r_1, \ldots, r_n)}(aM),$$

where $a > 0$ is a constant not depending on the function F_s.

REMARK. If it is assumed in addition to the conditions of Theorem [14.2] that

$$\prime = 1 - \frac{1}{p} \sum_{m+1}^{n} \frac{1}{r_i} > 0,$$

the sequence $\overset{\circ}{F}_s$ will converge to the function $\overset{\circ}{F}$.

For

$$\overset{\circ}{F}_s = \sum_{\nu=0}^{\infty} \overset{\circ}{q}_{\nu s} = \overset{\circ}{Q}_s + \sum_{\nu=s+1}^{\infty} \overset{\circ}{q}_{\nu s}, \ \overset{\circ}{F} = \overset{\circ}{Q}_s + \sum_{k=s+1}^{\infty} \left[\overset{\circ}{Q}_k - \overset{\circ}{Q}_{k-1}\right],$$

Therefore, using the same inequalities as in the proof of Theorem [14.1], we get

$$\left\| \overset{\circ}{F}_s - \overset{\circ}{F} \right\|_p^{(m)} = \left\| \sum_{\nu=s+1}^{\infty} \overset{\circ}{q}_{\nu s} - \sum_{k=s+1}^{\infty} \left(\overset{\circ}{Q}_k - \overset{\circ}{Q}_{k-1}\right) \right\|_p^{(m)}$$

$$\leqslant \sum_{\nu=s+1}^{\infty} \left\| \overset{\circ}{q}_{\nu s} \right\|_p^{(m)} + \sum_{k=s+1}^{\infty} \left\| \overset{\circ}{Q}_k - \overset{\circ}{Q}_{k-1} \right\|_p^{(m)} \leqslant \sum_{k=s+1}^{\infty} 2^n 2^{\frac{1}{p}\sum_{m+1}^{n} \frac{k}{r_i}} \left\| q_{\nu k} \right\|_p^{(n)}$$

$$+ \sum_{k=s+1}^{\infty} 2^n 2^{\frac{1}{p}\sum_{m+1}^{n} \frac{k}{r_i}} \left\| Q_k - Q_{k-1} \right\|_p^{(n)}$$

$$\leqslant 2^n cM \sum_{\nu=s+1}^{\infty} \frac{1}{2^{\varkappa\nu}} + 2^n \sum_{k=s+1}^{\infty} 2^{k(1-\varkappa)} \left[\left\| Q_k - F \right\|_p^{(n)} + \left\| Q_{k-1} - F \right\|_p^{(n)}\right]$$

$$\leqslant \frac{2^n cM}{2^{\varkappa s}} + 2^n \sum_{k=s+1}^{\infty} 2^{k(1-\varkappa)} \left(\frac{M(c+1)}{2^k} + \frac{M(c+1)}{2^{k-1}} \right)$$

$$\leqslant \frac{2^n cM}{2^{\varkappa s}} + \frac{2^{n+1}(c+1)M}{2^{\varkappa s}} \to 0, \quad \text{for } s \to \infty.$$

Q.E.D.

From this result we can obtain among other things a somewhat weakened form of Theorem [14.1]. To this end it suffices to extend all of the functions F_k converging in $L_p^{(n)}(G)$ in "the same way" from a subdomain Γ onto all of E^n with preservation of class so that the convergence of the sequence of functions F_s is preserved in $L_p^{(n)}(E^n)$. This can be done by means of the construction employed in the proof of Theorem [6.1].

[14.3]. *Suppose K is an $(n-1)$-dimensional manifold of smoothness $k \geqslant 2$, $K_{+\eta}^{(n)} = K_{+\eta}^{(n)}(K)$, $0 \leqslant \alpha < p-1$ and there is given on $K_{+\eta}^{(n)}$ a sequence of functions F_s converging in $W_p^{(1)}(r; \alpha; K_{+\eta}^{(n)})$: $\lim_{s\to\infty} F_s = F$, such that the sequence $F_s|_K$ converges in $L_p^{(n-1)}(K)$: $\lim_{s\to\infty} F_s|_K = f$. Then $F|_{\overset{\circ}{V}} = f$ on the base $\overset{\circ}{V}$ (see (8.6)) of any canonical neighborhood $V \subset K_\eta^{(n)}(K)$.*

Proof. We first note that by virtue of the convergence of the sequences F_s and $F_s|_K$ they are bounded in norm in the corresponding function spaces. Hence there exists a constant $M_1 > 0$ such that

$$\sum_{i=1}^{n}\left\|\frac{\partial F_s}{\partial x_i}\right\|^{(n)}_{\substack{p(r,\alpha)\\ K^{(n)}_{+\eta}}} < M_1, \quad \|F_s\|^{(n)}_{\substack{p\\ K^{(n)}_{+\eta}}} < M_1, \quad \|F_s|_K\|^{(n-1)}_p < M_1, \tag{14.10}$$

$$s=1,2\ldots,$$

Suppose now U is a canonical neighborhood of height ξ, $\overline{V}\subset U\subset K^{(n)}_{\eta_0}(K)$, $x(u)$ is the corresponding canonical mapping of U_u onto U and $U_{u\xi}=K^{(n-1)}_{+\xi}(\overset{\circ}{U}_u)$. Then the functions $F_s(x(u))$ are defined on $\overset{+}{U}_u$, and from (14.10) it follows that there exists a constant $M_2>0$ for which

$$\sum_{i=1}^{n}\left\|\frac{\partial F_s}{\partial u_i}\right\|^{(n)}_{\substack{p(u_n,\alpha)\\ \overset{+}{U}_u}} < M_2, \ \|F_s\|^{(n)}_{\substack{p\\ \overset{+}{U}_u}} < M_2, \ \|F_s\|^{(n-1)}_{\substack{p\\ \overset{\circ}{U}_u}} < M_2, \ s=1,2,\ldots .$$

Therefore the functions F_s satisfy the conditions of Theorem [11.14] for $\bar{r}=0$ and can consequently be extended onto U_u as functions $F^*_s(x(u))$ such that

$$F^*_s(x(u))\in H^{(r)}_p(M,\ U_u),\ r=1-\frac{\alpha}{p}>\frac{1}{p}, \tag{14.11}$$

where

$$M\leqslant A_1\sum_{i=1}^{n}\left\|\frac{\partial F_s}{\partial u_i}\right\|^{(n)}_{\substack{p(u_n,\alpha)\\ \overset{+}{U}_u}}+A_2\|F_s\|^{(n-1)}_{\substack{p\\ U_{u\xi}}}\leqslant A_1M_2+A_2\|F_s\|^{(n-1)}_{\substack{p\\ U_{u\xi}}},$$

But by Theorem [11.11]

$$\|F_s\|^{(n-1)}_{\substack{p\\ U_{u\xi}}}\leqslant\|F_s\|_{\overset{\circ}{U}_u}+B\sum_{i=1}^{n}\left\|\frac{\partial F_s}{\partial u_i}\right\|^{(n)}_{\substack{p(u_n,\alpha)\\ \overset{+}{U}_u}}\leqslant M_2+BM_2,$$

i.e., finally,

$$M\leqslant(A_1+A_2+A_2B)\,M_2. \tag{14.12}$$

On the other hand, if the extension is effected by means of the construction indicated in the proof of Theorem [11.14], we will have by virtue of (6.5)

$$\|F^*_s\|^{(n)}_{\substack{p\\ U_u}}\leqslant c\|F_s\|^{(n)}_{\substack{p\\ \overset{+}{U}_u}}\leqslant cM_2. \tag{14.13}$$

As is easily seen, owing to the completeness of the considered spaces, all that has been said about the functions F_s is also valid for F; in particular, it can also be extended onto U_u as a function F^*. Moreover, if we employ the indicated construction of an extension, we will have

$$\lim_{s\to\infty} F_s^* = F^* \tag{14.14}$$

in $L_p^{(n)}(U_u)$ and

$$\lim_{s\to\infty} F_s^*|_{\mathring{U}_u} = f \tag{14.15}$$

in $L_p^{(n-1)}(\mathring{U}_u)$. Finally, we extend the functions F_s^* from the domain V_u, which is the preimage of the given canonical neighborhood V under the mapping $x(u)$, onto all of the space according to Theorem [6.1]. It then follows from conditions (14.11)–(14.15) that the resulting sequence of functions satisfies all of the conditions of Theorem [14.1] for $V = G$. Therefore $F^*|_{\mathring{V}_u} = f$, and if we return to the manifold K by means of the canonical mapping $x(u)$, we will have $F|_{\mathring{V}} = f$. Q.E.D.

COMMENT. By making use of the Remarks to Theorem [14.2], Theorem [14.3] can be sharpened somewhat by dropping the requirement of convergence of the sequence $F_s|_K$; namely, it can be proved that the convergence of the $F_s|_{\mathring{V}}$ is a consequence of the convergence of the sequence F_s in $W_p^{(1)}(r; \alpha; K_{+\eta}^{(n)})$.

§15. Compact function classes

We now formulate the main compactness theorem. As in §14, let $\Phi(x_{m+1}, \cdots, x_n)$ denote the intersection of a set $\Phi \subset E^n$ with the hyperplane obtained by fixing the values of $x_{m+1}, \cdots, x_n$.

[15.1]. *Suppose* $1 \leqslant p \leqslant p' \leqslant \infty$, $1 \leqslant m \leqslant n$ *and*

$$\chi_m = 1 - \left(\frac{1}{p} - \frac{1}{p'}\right)\sum_{j=1}^{m}\frac{1}{r_j} - \frac{1}{p}\sum_{i=m+1}^{n}\frac{1}{r_i}, \quad \omega_\tau = \chi_m - \sum_{j=1}^{m}\frac{\tau_j}{r_j},$$

$$\chi_\lambda = \chi_m - \sum_{k=m+1}^{n}\frac{\lambda_k}{r_k}.$$

Then it is possible to extract from every bounded (in the sense of the norm of $L_p^{(n)}$*) system* $\mathfrak{S} = \{f\}$ *of functions of class* $H_p^{(r_1,\cdots,r_n)}(M_1, \cdots, M_n)$, $\|f\|_p^{(n)} \leqslant M$, $f \in \mathfrak{S}$, *a sequence* $f_k \in \mathfrak{S}$, $k = 1, 2, \cdots$, *such that for any fixed values* $x_{m+1}^{(0)}, \cdots, x_n^{(0)}$ *and compact set* $\Phi \subset E^n$ *the sequence*

$$f_k(x_1, \ldots, x_m, x_{m+1}^{(0)}, \ldots, x_n^{(0)})$$

together with all of its derivatives $f^{(\tau)}_{x_1^{\tau_1}\ldots x_m^{\tau_m}}$ *and normal derivatives* $f^{(\lambda)}_{x_{m+1}^{\lambda_{m+1}}\ldots x_n^{\lambda_n}}$, *for which* $\omega_\tau > 0$ *and* $\chi_\lambda > 0$ *respectively converges in* $L_p^{(m)}(\Phi(x_{m+1}^{(0)}, \cdots, x_n^{(0)})$, *this convergence being uniform relative to* $x_{m+1}^{(0)}, \cdots, x_n^{(0)}$. *Further, the limits of the normal derivatives are the corresponding normal derivatives of the limit function and, for any bounded domain* $G \subset E^m$,

$$F^{(\tau)}_{x_1^{\tau_1}\ldots x_m^{\tau_m}} \in H_{p'}^{(\sigma_1, \ldots, \sigma_m)}(M_\tau, G),$$

where $\sigma_j = r_j\omega_\tau$, *while*

$$F^{(\lambda)}_{x_{m+1}^{\lambda_{m+1}}\ldots x_n^{\lambda_n}} \in H_{p'}^{(\beta_1, \ldots, \beta_m)}(M_\lambda, G),$$

where $\beta_j = r_j\chi_\lambda, j = 1, \cdots m$, *the constants* M_τ *and* M_λ *being linear combinations of the constants* $M, M_1, \cdots, M_n$.

This theorem was proved by the author in [15] and is a generalization, as has already been mentioned in the Introduction, of the corresponding theorem of Nikol'skiĭ.

It can easily be carried over to the case of arbitrary manifolds by the usual method of canonical neighborhoods and canonical mappings (see, for example, the same paper [15]). Here we merely give some applications of it to spaces of functions whose derivatives are summable in the pth power with power weight.

Suppose $(x_i) \in E^n$, Γ is a finite or infinite domain in the hyperplane $E^m = \{(x_i): x_{m+1} = \cdots = x_n = 0\}$, possibly coinciding with E^m itself,

$$0 < \eta < a < \infty, \quad \Gamma_\eta = \{P: P \in \Gamma, \ \rho(P, E^m \setminus \Gamma) > \eta\},$$
$$G = \{(x_i): (x_1, \ldots, x_m) \in \Gamma, |x_j| < a, j = m+1, \ldots, n\},$$
$$G_\eta = \{(x_i): (x_1, \ldots, x_m) \in \Gamma_\eta, |x_j| < a - \eta, j = m+1, \ldots, n\},$$
$$G(x_j^{(0)}) = G \cap \{(x_i): x_j = x_j^{(0)}\}, \ \Phi(x_j^{(0)}) = \Phi \cap \{(x_i): x_j = x_j^{(0)}\},$$
$$j = m+1, \ldots, n \text{ and } r = \sqrt{x_{m+1}^2 + \ldots + x_n^2}.$$

[15.2]. *Suppose* $1 \leqslant p < \infty, 0 \leqslant \alpha < p-1$ *and* $\mathfrak{S}$ *is a system of functions* f *that together with their weak derivatives of first order are defined on* G *and bounded in* $W_p^{(1)}(r; 0, \alpha; G)$, *i.e. there exists a constant* $M > 0$ *for which*

$$\|f\|^{(n)}_{W^{(1)}_p(r;\,0,\,\alpha;\,G)} \leqslant M, \ f \in \mathfrak{S}. \tag{15.1}$$

Then there exists a sequence $f_k \in \mathfrak{S}$ $(k = 1, 2, \cdots)$ *such that for any compact set* $\Phi \subset G$ *the following conditions hold*:

1. *The sequence* f_k *converges in* $L_p^{(n)}(\Phi)$.

2. *For any fixed* $x_j^{(0)}$ *the sequence* $f_k|_{x_j = x_j^{(0)}}$ *converges in* $L_p^{(n-1)}(\Phi(x_j^{(0)}))$ *and, moreover, uniformly relative to the choice of* $x_j^{(0)}, j = m+1, \cdots, n$.

PROOF. Suppose first the value of $j = m+1, \cdots, n$ is fixed. From condition (15.1) it follows, in particular, that

$$\|f\|^{(n)}_{p,\,G} \leqslant M. \tag{15.2}$$

There therefore obviously exists an $x_j^{(0)} \in [0, a]$ such that

$$\|f\|^{(n-1)}_{p,\,\overline{G}(x_j^{(0)})} \leqslant \frac{M}{a}. \tag{15.3}$$

For suppose the contrary; for all $x_j \in [0, a]$

$$\|f\|^{(n-1)}_{p,\,\overline{G}(x_j)} > \frac{M}{a}.$$

Then

$$\|f\|^{(n)}_{p,\,G} \geqslant \left\{\int_0^a dx_j \int \cdots \int_{\overline{G}(x_j)} |f(x_1, \ldots, x_n)|^p \, dx_1 \ldots dx_{j-1}\, dx_{j+1} \ldots dx_n\right\}^{\frac{1}{p}} \geqslant 2M,$$

which contradicts (15.2). From (15.3), using [11.11],(8) we get

$$\|f\|^{(n-1)}_{p,\,\overline{G}(a)} \leqslant \|f\|^{(n-1)}_{p,\,\overline{G}(x_j^0)} + 2(a+1)B\|f_{x_j}\|^{(n)}_{p(x_j,\alpha),\,G} \leqslant \|f\|^{(n-1)}_{p,\,\overline{G}(x_j^0)} + 2(a+1)B\|f_{x_j}\|^{(n)}_{p(r,\alpha),\,G}$$
$$\leqslant \frac{M}{a} + 2(a+1)BM. \tag{15.4}$$

It is analogously proved that

$$\|f\|^{(n-1)}_{p,\,G(-a)} \leqslant \frac{M}{a} + 2(a+1)BM. \tag{15.4'}$$

---

(8)Under our assumptions $a^{(p-1-\alpha)/p} < a+1$.

According to Theorem [11.7] we have

$$f \in H_p^{(r)}(M_1, G), \quad r = 1 - \frac{\alpha}{p} > \frac{1}{p}, \tag{15.5}$$

where

$$M_1 \leqslant A_1 \left[\| f \|_{p \atop G(a)}^{(n-1)} + \| f \|_{p \atop G(-a)}^{(n-1)} \right] + B_1 \sum_{i=1}^{n} \| f_{x_i} \|_{p(r,\alpha) \atop G}^{(n)}$$
$$\leqslant M \left(\frac{2A_1}{a} + 4A_1(a+1)B + B_1 \right). \tag{15.6}$$

We now choose for each fixed compact set Φ a number $\eta > 0$ in such a way that $\Phi \subset G_\eta$ and extend each function $f \in \mathfrak{S}$ from G_η to all of E^n according to Theorem [6.1]. Then by virtue of conditions (15.2), (15.5) and (15.6) the resulting system of extensions of the functions of $\mathfrak{S}$ will satisfy all of the requirements of [15.1], and hence a sequence can be extracted from it for which propositions 1° and 2° of the theorem being proved hold on the given compact set Φ, on which the extensions coincide with the original functions. To obtain the desired sequence it suffices to choose a sequence of numbers $\eta_\nu \to 0, \nu = 1, 2, \cdots$ and apply the diagonal process relative to $\nu = 1, 2, \cdots$ and $j = m+1, \cdots, n$ to the construction described above.

In conclusion we note a compactness criterion for halfspaces.

[15.3]. *Suppose* $1 \leqslant m < \infty, 0 \leqslant \alpha < p - 1, E^{n-1} = \{(x_i): x_n = 0\}$ *and* $\mathfrak{S}$ *is a system of functions* f *that together with their weak derivatives of first order are defined on the halfspace* $\overset{+}{E}{}^n = \{(x_i): x_n \geqslant 0\}$ *and satisfy the conditions*

$$\| f \|_{p \atop E^{n-1}}^{(n-1)} \leqslant M, \quad \| f_{x_i} \|_{p(x_n,\alpha) \atop \overset{+}{E}{}^n}^{(n)} \leqslant M, \quad i = 1, 2, \ldots, n, \; f \in \mathfrak{S},$$

in which M *is a constant. Then there exists a sequence of functions* $f_k \in \mathfrak{S}$ *such that for any compact set* $\Phi \subset \overset{+}{E}{}^n$

1°. *the sequence* f_k *converges in* $L_p^{(n)}(\Phi)$, *and*

2°. *the sequence* $f_k|_{x_n = x_n^{(0)}}$ *converges in* $L_p^{(n-1)}(\Phi(x_n^{(0)}))$ *for any* $x_n^{(0)}$ *and, moreover, uniformly relative to the choice of* $x_n^{(0)}$.

The proof of this theorem is essentially analogous to the proof of the preceding theorem if one takes into account that each function can be extended onto all of the space according to Theorem [11.14].

These theorems can be carried over to the case $p = \infty$ by the usual method.

§16. Weighted imbedding theorems for domains with a piecewise smooth boundary

In the present section we confine ourselves to the consideration of a finite domain $G \subset \overset{+}{E}{}^n = \{(x_i): x_n > 0\}$, satisfying the conditions of §9, and we will use, as a rule, all of the notation introduced there without additional references. Let $h > 0$, $\overset{+}{E}{}^n_{(h)} = \{(x_i): 0 < x_n < h\}$, $G_{(h)} = G \cap \overset{+}{E}{}^n_{(h)}$ and $\overset{+}{G}_h = G \setminus \overline{G}_{(h)}$. We will assume that there exists an $h > 0$ such that $\overset{+}{G}_h$ is regular relative to its boundary (see Definition [12.7]), which contains all of the manifolds M_μ (see (9.1)). Let $\mathring{G}_{(h)}$ denote the projection of $G_{(h)}$ onto the hyperplane $E^{n-1} = \{(x_i): x_n = 0\}$. We will assume in addition that the intersection with $G_{(h)}$ of every straight line parallel to the Ox_n axis and passing through a point $(x_1, \cdots, x_{n-1}) \in \mathring{G}_{(h)}$ is an interval. This restriction is in no way connected with the essence of the question being studied; but it permits one to simplify quite substantially the technical aspect of some of the proofs without changing them in principle.

If K is the boundary of G, a point $P \in K \setminus (\overline{K}^+ \cap E^{n-1})$ will be called a *proper point* (it is not a corner point), and a neighborhood of it in K will be denoted by $U_K(P)$.

[16.1]. *Suppose given on G a function f together with its weak derivatives of first order such that*

$$\|f_{x_i}\|^{(n)}_{\substack{p(x_n,\alpha)\\G}} < \infty, \quad i = 1, 2, \ldots, n, \quad 0 \leqslant \alpha < p-1, \quad 1 \leqslant p < \infty.$$

Then for any proper point $P \in K$ there exists a neighborhood $U_K(P)$ of it such that the boundary value $f|_{U_K(P)}$ exists in the sense of a mean value and satisfies the condition

$$f|_{U_K(P)} \in H_p^{(r)}(U_K(P)), \quad r = 1 - \frac{\alpha}{p}.$$

This immediately follows from Theorem [12.3] for $\lambda = 1$. If $P \notin E^{n-1}$, it can even be asserted that $r = \frac{1}{2}$.

REMARK. In the case under consideration one can always speak of the value $f|_K$ of f on the boundary K of G since the boundary points that are not proper points form a set of measure zero on the boundary (see §8). We could have shown that $f|_K \in L_p^{(m-1)}(K)$, but the proof is quite tedious. Besides, it follows at once if one assumes that the part K^+ of the boundary intersects the hyperplane E^{n-1} at an angle greater than $\pi/2$. In this case an $\eta > 0$ can be chosen so large that the closures of the manifold pieces $K_{+\eta}^{(n-1)}(K_\kappa)$ and $K_{+\eta}^{(n-1)}(L_\lambda)$ (see (8.9) and (9.1)) are entirely contained in the interior of G. Then by virtue of the same Theorem [12.3] for $\alpha = 0$ or, correspondingly, a theorem of Nikol'skiĭ ([4], page 104) the function f is p power summable. After this it suffices to estimate the norm of f on the manifolds K_κ and L_λ, $\kappa = 1, \cdots, \kappa_0$, $\lambda = 1, \cdots, \lambda_0$, with the use of Theorem [12.6].

If K^+ does not intersect E^{n-1} at a zero angle, one can extend f in a neighborhood of $K^+ \cap E^{n-1}$ from $G_{(h)}$ (for sufficiently small h) to a domain whose boundary intersects the hyperplane E^{n-1} at an angle greater than $\pi/2$ and thereby reduce this case to the preceding one. We will not explicitly reproduce this proof, since it suffices for our purposes that the function $f|_K$ exist on the boundary K of G and locally belong to the corresponding H-class. In those cases when the norm of a function on the boundary plays a role its finiteness will always be included in the assumptions of the theorem.

[16.2]. *Suppose K is the boundary of the domain G, $1 \leqslant p < \infty$, $0 \leqslant \alpha < p - 1$ and a function f together with its weak derivatives of first order is defined on G, the norm $\|f_{x_i}\|^{(n)}_{p(x_n,\alpha)}$ (over G), $i = 1, \cdots, n$, being finite. Then there exist constants $c_1 > 0$ and $c_2 > 0$ not depending on f such that*

$$\underset{G}{\|f\|_p^{(n)}} \leqslant c_1 \underset{K}{\|f\|_p^{(n-1)}} + c_2 \sum_{i=1}^{n} \underset{G}{\|f_{x_i}\|^{(n)}_{p(x_n,\alpha)}}.$$

COROLLARY. *In the case $\alpha = 0$, $p = 2$ this result implies the well-known inequality* (*see* [2], *page* 89)

$$\underset{G}{\|f\|_2^{(n)}} \leqslant c_1 \underset{K}{\|f\|_2^{(n-1)}} + c_2 \{D(f)\}^{\frac{1}{2}}.$$

The proof of this theorem is analogous to the proof of Theorem [12.8] and slightly differs from it only in regard to the characteristics of the domain G. For each point $(x_1, \cdots, x_{n-1}) \in \overset{\circ}{G}_{(h)}$, where h is fixed (see the introductory remarks), we denote by $\psi = \psi(x_1, \cdots, x_{n-1})$ and $\varphi = \varphi(x_1, \cdots, x_{n-1})$ the nth coordinate of the upper and lower ends respectively of the interval of intersection of a straight line parallel to the Ox_n axis with the domain $G_{(h)}$. Clearly, $0 \leqslant \psi - \varphi \leqslant h$ on $\overset{\circ}{G}_{(h)}$.

Analogously to Theorem [11.2], we then have

$$\|f\|_{p,G_{(h)}}^{(n)} = \Big\{\int\!\!\cdots\!\!\int_{G_{(h)}} \Big| f(x_1, \ldots, x_{n-1}, \varphi) + \int_{\varphi}^{x_n} f_{x_n}(x_1, \ldots, x_{n-1}, t)\Big|^p dx_1 \ldots dx_n\Big\}^{\frac{1}{p}}$$

$$\leqslant \Big\{\int\!\!\cdots\!\!\int_{\overset{\circ}{G}_{(h)}} \big| f(x_1, \ldots, x_{n-1}, \varphi)\big|^p dx_1 \ldots dx_{n-1} \int_{\varphi}^{\psi} dx_n\Big\}^{\frac{1}{p}}$$

$$+ \Big\{\int\!\!\cdots\!\!\int_{G_{(h)}} \Big|\int_{\varphi}^{x_n} f_{x_n}(x_1, \ldots, x_{n-1}, t)\, dt\Big|^p dx_1 \ldots dx_n\Big\}^{\frac{1}{p}}$$

$$\leqslant h^{\frac{1}{p}} \Big\{\int\!\!\cdots\!\!\int_{\overset{\circ}{G}_{(h)}} \big| f(x_1, \ldots, x_{n-1}, \varphi)\big|^p dx_1 \ldots dx_{n-1}\Big\}^{\frac{1}{p}}$$

$$+ \Big\{\int\!\!\cdots\!\!\int_{G_{(h)}} \int_{\varphi}^{x_n} t^{\alpha} \big| f_{x_n}(x_1, \ldots, x_{n-1}, t)\big|^p dt \left(\int_{\varphi}^{x_n} \frac{dt}{t^{\frac{\alpha q}{p}}}\right)^{\frac{p}{q}} dx_1 \ldots dx_{n-1}\, dx_n\Big\}^{\frac{1}{p}}$$

$$\leqslant h^{\frac{1}{p}} \|f\|_{p,K}^{(n-1)} + \left(\frac{p}{p-1-\alpha}\right)^{\frac{1}{q}} h^{1-\alpha} \|f_{x_n}\|_{p(x_n,\alpha),G}^{(n)}. \tag{16.1}$$

We denote the boundary of the domain $\overset{+}{G}_h$ by K_h. By virtue of the fact that $\overset{+}{G}_h$ is regular relative to its boundary we can partition it into subdomains $G^{(\nu)}$, and its boundary K_h into places $\Gamma^{(\nu)}$, $\nu = 1, \cdots, \nu_0$, such that (see (12.13))

$$\|f\|_{p,G^{(\nu)}}^{(n)} \leqslant d^{\frac{1}{p}} \|f\|_{p,\Gamma^{(\nu)}}^{(n-1)} + d\|f_{x_i}\|_{p,G^{(\nu)}}^{(n)},$$

where i as a function of ν can take the values $1, \cdots, n$ and d is the diameter of G. Further,

$$\|f_{x_i}\|_{p,G^{(\nu)}}^{(n)} \leqslant \Big\{\int\!\!\cdots\!\!\int_{G^{(\nu)}} \left(\frac{x_n}{h}\right)^{\alpha} |f_{x_i}|^p\, dx_1 \ldots dx_n\Big\}^{\frac{1}{p}} \leqslant h^{-\frac{\alpha}{p}} \|f_{x_i}\|_{p(x_n,\alpha),G}^{(n)},$$

which implies

$$\|f\|_{p,\overset{+}{G}_h}^{(n)} \leqslant \sum_{\nu=1}^{\nu_0} \|f\|_{p,G^{(\nu)}}^{(n)} \leqslant \nu_0 d^{\frac{1}{p}} \|f\|_{p,K_h}^{(n-1)} + \nu_0 d h^{-\frac{\alpha}{p}} \sum_{i=1}^{n} \|f_{x_i}\|_{p(x_n,\alpha),G}^{(n)}. \tag{16.2}$$

Let $\overset{\circ}{K}_h = K_h \cap \{(x_i): x_n = h\}$. Then $K_h \subset K \cup K_D$, and hence

$$\|f\|_{p\,K_h}^{(n-1)} \leqslant \|f\|_{p\,K}^{(n-1)} + \|f\|_{p\,\mathring{K}_h}^{(n-1)}.$$

But

$$\|f\|_{p\,\mathring{K}_h}^{(n-1)} \leqslant \Big\{\int \cdots \int_{\mathring{G}_{(h)}} |f(x_1, \ldots, x_{n-1}, \psi)|^p \, dx_1 \ldots dx_{n-1}\Big\}^{\frac{1}{p}}$$

$$\leqslant \Big\{\int \cdots \int_{\mathring{G}_{(h)}} \Big|f(x_1, \ldots, x_{n-1}, \varphi) + \int_{\varphi}^{\psi} f_{x_n}(x_1, \ldots, x_{n-1}, t)\, dt\Big|^p dx_1 \ldots dx_{n-1}\Big\}^{\frac{1}{p}}$$

$$\leqslant \|f\|_{p\,K}^{(n-1)} + B' \|f_{x_n}\|_{p(x_n,\alpha)\,G_{(h)}}^{(n)} \leqslant \|f\|_{p\,K}^{(n-1)} + B' \|f_{x_n}\|_{p(x_n,\alpha)\,G}^{(n)},$$

where B' is a constant (cf. [11.11]). Thus

$$\|f\|_{p\,K_h}^{(n-1)} \leqslant 2\|f\|_{p\,K}^{(n-1)} + B' \|f_{x_n}\|_{p(x_n,\alpha)\,G}^{(n)}.$$

Therefore (16.2) implies

$$\|f\|_{p\,\overset{+}{G}_h}^{(n)} \leqslant 2\nu_0 d^{\frac{1}{p}} \|f\|_{p\,K}^{(n-1)} + \Big(B' + \nu_0 d h^{-\frac{\alpha}{p}}\Big) \sum_{i=1}^{n} \|f_{x_i}\|_{p(x_n,\alpha)\,G}^{(n)}. \tag{16.3}$$

Finally, noting that

$$\|f\|_{p\,G}^{(n)} \leqslant \|f\|_{p\,G_{(h)}}^{(n)} + \|f\|_{p\,\overset{+}{G}_h}^{(n)},$$

we obtain the desired inequality from inequalities (16.1) and (16.3).

We conclude this section by noting that, analogously to §13, a complete normed linear space $W_p^{(r)}(x_n; \alpha_0, \cdots, \alpha_r; G)$ can be introduced in a natural way for the domain G.

For example, when $r = 1, \alpha_0 = 0, \alpha_1 = \alpha$

$$\|f\|_{W_p^{(1)}(x_n; 0, \alpha; G)}^{(n)} = \|f\|_{p\,G}^{(n)} + \sum_{i=1}^{n} \|f_{x_i}\|_{p\,G}^{(n)}(x_n, \alpha). \tag{16.4}$$

Therefore, if a function f satisfies the conditions

$$\underset{K}{\|f\|_p^{(n-1)}} < \infty, \sum_{i=1}^{n} \underset{G}{\|f_{x_i}\|_{p(x_n,\alpha)}^{(n)}} < \infty, \ 0 \leqslant \alpha < p-1, \tag{16.5}$$

Theorem [16.2] implies

$$f \in W_p^{(1)}(x_n;\, 0,\, \alpha;\, G). \tag{16.6}$$

THIRD CHAPTER

VARIATIONAL METHODS FOR THE SOLUTION OF ELLIPTIC EQUATIONS

§17. General plan of the variational method

Let G be a bounded domain (or even a bounded open set) of n-dimensional space E^n, and let Γ be its boundary. By $L_2 = L_2(G)$ we denote the Hilbert space of functions defined and square summable on G. For two functions $u \in L_2$ and $v \in L_2$ the symbol (u, v) will denote, as usual, the scalar product

$$(u, v) = \int uv\,dG$$

and for $u \in L_2$ we write

$$\|u\|_{L_2} = \sqrt{(u, u)}.$$

By $W_2^{(1)} = W_2^{(1)}(G)$ we denote the space of functions having weak derivatives of first order on G which together with the function itself belong to L_2. For the norm of a function $u \in W_2^{(1)}$ we take

$$\|u\|_{W_2^{(1)}} = \|u\|_{L_2} + \sum_{i=1}^{n} \left\| \frac{\partial u}{\partial x_i} \right\|_{L_2}. \tag{17.1}$$

As is well known [2, 4], the space $W_2^{(1)}$ is complete. Further, for $u \in W_2^{(1)}$ and $v \in W_2^{(1)}$ we write

$$D(u, v) = \int \sum_{i=1}^{n} \frac{\partial u}{\partial x_i} \frac{\partial v}{\partial x_i} dG, \quad D(u) = D(u, u).$$

It is easily seen that every function $u \in W_2^{(1)}$ satisfies the inequalities

$$D^{\frac{1}{2}}(u) \leqslant \sum_{i=1}^{n} \left\| \frac{\partial u}{\partial x_i} \right\|_{L_2} \leqslant n^{\frac{1}{2}} D^{\frac{1}{2}}(u) \tag{17.2}$$

(the right inequality follows from Hölder's inequality).

Let $A_{ik} = A_{ik}(x_1, \cdots, x_n)$, $B_i = B_i(x_1, \cdots, x_n)$ and $C = C(x_1, \cdots, x_n)$ be bounded measurable functions defined on G, and let u and v be elements of $W_2^{(1)}$. We consider the bilinear functional

$$A(u, v) = \int \left[\sum_{i=1}^{n} \sum_{k=1}^{n} A_{ik} \frac{\partial u}{\partial x_i} \frac{\partial v}{\partial x_k} + \sum_{i=1}^{n} B_i \left(\frac{\partial u}{\partial x_i} v + u \frac{\partial v}{\partial x_i} \right) + Cuv \right] dG \qquad (17.3)$$

and the corresponding "quadratic form" $A(u) = A(u, u)$. It can be assumed without loss of generality for the problems connected with the study of $A(u)$ below that $A_{ik} = A_{ki}$ $(i, k = 1, \cdots, n)$. For otherwise one should introduce in the expression for $A(u)$ the new coefficients $A_{ik}^* = A_{ki}^* = (A_{ik} + A_{ki})/2$.

We will assume that $A(u)$ is positive definite, i.e. that there exists a constant $\beta > 0$ such that the following inequality (ellipticity condition) is satisfied:

$$A(u) \geqslant \beta D(u);^{(1)} \qquad (17.4)$$

in particular

$$A(u) \geqslant 0. \qquad (17.5)$$

On the other hand it is easily seen that there exists a $\gamma > 0$ such that

$$A^{1/2}(u) \leqslant \gamma \|u\|_{W_2^{(1)}} \qquad (17.6)$$

for any function $u \in W_2^{(1)}$.

We note that $A(u, v)$, like every nonnegative symmetric bilinear functional, satisfies the inequality

$$A(u, v) \leqslant A^{1/2}(u) A^{1/2}(v) \qquad (17.7)$$

as well as the triangle inequality

$$A^{1/2}(u + v) \leqslant A^{1/2}(u) + A^{1/2}(v), \qquad (17.8)$$

and hence

$$|A^{1/2}(u) - A^{1/2}(v)| \leqslant A^{1/2}(u - v), \qquad (17.9)$$

---

(1) For this to be true it suffices, for example, that the following inequality hold for any real $\xi, \xi_1, \cdots, \xi_n$:

$$\sum_{i,k=1}^{n} A_{ik} \xi_i \xi_k + 2 \sum_{i=1}^{n} B_i \xi_i \xi + C\xi^2 \geqslant \beta \sum_{i=1}^{n} \xi_i^2.$$

where $u \in W_2^{(1)}, v \in W_2^{(1)}$.

As is well known, for any function $u \in W_2^{(1)}$ (see the author's paper [16] as well as the Corollary to [11.3] in the present paper) there exist determinant boundary values in the sense of convergence almost everywhere in a direction parallel to any of the coordinate axes (or, of course, to any other fixed direction).

More precisely, this means the following. Let $\mathrm{pr}_i\, G$ denote the orthogonal projection of G onto the hyperplane $x_i = 0$. It is obvious that for any $i = 1, \cdots, n$ and any point $x \in \mathrm{pr}_i\, G$ the intersection of G with the straight line passing through x and perpendicular to the hyperplane $x_i = 0$ is composed of not more than a countable number of intervals, which we denote, after an arbitrary enumeration, as follows:

$$\left(a_{k_x}^{(i)}, b_{k_x}^{(i)}\right), \quad k_x = 1, 2, \ldots, x \in \mathrm{pr}_i G.$$

We always assume $a_{k_x}^{(i)} < b_{k_x}^{(i)}$. The set of all $a_{k_x}^{(i)}$ $(b_{k_x}^{(i)})$ will be called the ith lower (upper) halfboundary of G. Now for fixed $i = 1, \cdots, n$ every function $u \in W_2^{(1)}$ can be modified on a set of measure zero so that the limits

$$\lim_{x \to a_{k_x}^{(i)}+0} u(x) \text{ and } \lim_{x \to b_{k_x}^{(i)}-0} u(x),$$

exist for almost all $x \in \mathrm{pr}_i\, G$ and all $k_x = 1, 2, \cdots$ and can therefore be taken as boundary values at the corresponding points of the boundary. It turns out that for almost all $x \in \mathrm{pr}_i\, G$ (in the sense of an $(n-1)$-dimensional measure) these limits do not depend on the indicated modification of the function u on a set of n-dimensional measure zero.

Suppose Γ is the boundary of a domain G and $\gamma \in \Gamma$, almost all (in the above indicated sense) of the points of γ being accessible in the direction of the ith coordinate axis. Then the boundary values on γ corresponding to this direction of a function $u \in W_2^{(1)}(G)$ will be denoted by $u|_\gamma^{(i)}$. If a function φ is given on Γ (more precisely, on the set of its points that are accessible in at least one coordinate direction) which is the boundary function in our sense of a function $u \in W_2^{(1)}$ with respect to all of the coordinate directions, we will simply write $u|_\Gamma = \varphi$.

It is shown in the author's paper [16] that if i is fixed $(i = 1, \cdots, n)$, $\gamma \in \Gamma$, γ contains one of the ith halfboundaries of G, $u \in W_2^{(1)}$ and $u|_\gamma^{(i)} = 0$, then

$$(u, u) \leqslant \delta^2 D(u), \tag{17.10}$$

where δ is the diameter of G $(\delta > 0)$.(2)

Suppose given on G another function $f \in L_2$ and let

$$K(u) = A(u) - 2(f, u). \tag{17.11}$$

Further, suppose given on the above-considered part γ of the boundary of G a function φ, and let $\mathfrak{K}_\varphi^{(i)} = \mathfrak{K}_\varphi^{(i)}(\gamma)$ denote the set of all functions $u \in W_2^{(1)}$ such that $u|_\gamma^{(i)} = \varphi$.

Suppose the set $\mathfrak{K}_\varphi^{(i)}$ is not empty, i.e. that there exists at least one function $v_0 \in \mathfrak{K}_\varphi^{(i)}$. Then the functional $K(u)$ is bounded from below on $\mathfrak{K}_\varphi^{(i)}$. For if $u \in \mathfrak{K}_\varphi^{(i)}$, then

$$\|u\|_{L_2} - \|v_0\|_{L_2} \leqslant \|u - v_0\|_{L_2} \leqslant \delta D^{\frac{1}{2}}(u - v_0) \leqslant \delta D^{\frac{1}{2}}(u) + \delta D^{\frac{1}{2}}(v_0),$$

and hence

$$(u, u) \leqslant 3\delta^2 D(u) + 3\delta^2 D(v_0) + 3(v_0, v_0).$$

Therefore

$$\begin{aligned} K(u) &= A(u) - 2(f, u) \\ &\geqslant A(u) - \left[\frac{6\delta^2}{\beta}(f, f) + \frac{\beta}{6\delta^2}(u, u)\right] \\ &\geqslant A(u) - \left[\frac{6\delta^2}{\beta}(f, f) + \frac{\beta}{2} D(u) + \frac{\beta}{2} D(v_0) + \frac{\beta}{2\delta^2}(v_0, v_0)\right] \\ &\geqslant A(u) - \left[\frac{6\delta^2}{\beta}(f, f) + \frac{1}{2} A(u) + \frac{\beta}{2} D(v_0) + \frac{\beta}{2\delta^2}(v_0, v_0)\right] \\ &\geqslant \frac{1}{2} A(u) - \left[\frac{6\delta^2}{\beta}(f, f) + \frac{\beta}{2} D(v_0) + \frac{\beta}{2\delta^2}(v_0, v_0)\right] \\ &\geqslant -\left[\frac{6\delta^2}{\beta}(f, f) + \frac{\beta}{2} D(v_0) + \frac{\beta}{2\delta^2}(v_0, v_0)\right]. \end{aligned} \tag{17.12}$$

We show that there exists a function $u_0 \in \mathfrak{K}_\varphi^{(i)}$ such that

$$K(u_0) = \min_{u \in \mathfrak{K}_\varphi^{(i)}} K(u).$$

---

(2) This inequality together with all of the consequent assertions below remains valid if γ has the above described property merely with respect to a curvilinear coordinate system. Thus, if Γ is a circle, any arc can be taken as γ.

Let

$$k = \min_{u \in \mathfrak{R}_{\varphi}^{(i)}} K(u).$$

We fix a sequence $u_m \in \mathfrak{R}_{\varphi}^{(i)}$ such that $\lim_{m \to \infty} K(u_m) = k$, and take $\epsilon > 0$. We next choose a number m_ϵ so that $K(u_{m+\mu}) < k + \epsilon$ for $m \geqslant m_\epsilon$ and $\mu = 0$, $1, 2, \cdots$. Then, since $(u_{m+\mu} + u_m)/2 \in \mathfrak{R}_{\varphi}^{(i)}$, we have $K((u_{m+\mu} + u_m)/2) \geqslant k$. Further, verifying by a direct calculation that the identity

$$A\left(\frac{u-v}{2}\right) = \frac{1}{2} K(u) + \frac{1}{2} K(v) - K\left(\frac{u+v}{2}\right) \tag{17.13}$$

holds for any functions $u \in W_2^{(1)}$ and $v \in W_2^{(1)}$, we get

$$A\left(\frac{u_{m+\mu} - u_m}{2}\right) = \frac{1}{2} K(u_m) + \frac{1}{2} K(u_{m+\mu}) - K\left(\frac{u_{m+\mu} + u_m}{2}\right)$$
$$\leqslant \frac{k+\varepsilon}{2} + \frac{k+\varepsilon}{2} - k = \varepsilon.$$

Hence

$$A(u_{m+\mu} - u_m) < 4\epsilon.$$

This implies by virtue of inequalities (17.4) and (17.10) that for $m \geqslant m_\epsilon$ and $\mu = 0, 1, \cdots$

$$D(u_{m+\mu} - u_m) < \frac{4\varepsilon}{\beta},$$
$$\| u_{m+\mu} - u_m \|_{L_2} \leqslant 2\delta \sqrt{\frac{\varepsilon}{\beta}}.$$

It follows by virtue of (17.2) and (17.1) that $\{u_m\}$ is a Cauchy sequence in $W_2^{(1)}$, and in view of the completeness of the latter (see, for example, [2]) there exists a function $u_0 \in W_2^{(1)}$ such that

$$\lim_{m \to \infty} \| u_m - u_0 \|_{W_2^{(1)}} = 0.$$

It was shown in the author's paper [16] that the limit of the boundary values of a sequence of functions converging in $W_2^{(1)}$ coincides almost everywhere with the boundary values of the limit function. Therefore $u_0 |_{\Gamma}^{(i)} = \varphi$, and hence $u_0 \in \mathfrak{R}_{\varphi}^{(i)}$.

Let us prove that $K(u_0) = k$. From inequalities (17.6) and (17.9) we have

$$|K(u_m)-K(u_0)|\leqslant|A(u_m)-A(u_0)|+2|(f,u_m-u_0)|$$
$$\leqslant|A^{\frac{1}{2}}(u_m)-A^{\frac{1}{2}}(u_0)|\,[A^{\frac{1}{2}}(u_m)+A^{\frac{1}{2}}(u_0)]$$
$$+2\|f\|_{L_2}\|u_m-u_0\|_{L_2}\leqslant A^{\frac{1}{2}}(u_m-u_0)[A^{\frac{1}{2}}(u_m)+A^{\frac{1}{2}}(u_0)]$$
$$+2\|f\|_{L_2}\|u_m-u_0\|_{W_2^{(1)}}\leqslant\gamma\|u_m-u_0\|_{W_2^{(1)}}\left[A^{\frac{1}{2}}(u_m)+A^{\frac{1}{2}}(u_0)+\frac{2\|f\|_{L_2}}{\gamma}\right].$$

This implies that

$$k=\lim_{m\to\infty}K(u_m)=K(u_0).$$

It is not difficult to see that the function effecting a minimum of the functional $K(u)$ in the class $\mathfrak{K}_\varphi^{(i)}$ is unique. For suppose $u\in\mathfrak{K}_\varphi^{(i)}$ and $K(u)=k$. Applying (17.13), we get

$$0\leqslant A\left(\frac{u-u_0}{2}\right)=\frac{1}{2}K(u)+\frac{1}{2}K(u_0)-K\left(\frac{u+u_0}{2}\right)\leqslant\frac{k}{2}+\frac{k}{2}-k=0,$$

i.e. $A(u-u_0)=0$. But then $D(u-u_0)=0$ by virtue of inequality (17.4). Therefore, since $(u-u_0)|_\gamma^{(i)}=0$, according to (17.10) we have $\|u-u_0\|_{L_2}=0$, which implies that u coincides with u_0 as an element of L_2 and hence as an element of $W_2^{(1)}$.

We have thus proved the following theorem.

[17.1]. *Suppose γ is a subset of the boundary of a domain G containing one of the ith halfboundaries of G, φ is a function defined on γ, the class $\mathfrak{K}_\varphi^{(i)}=K_\varphi^{(i)}(\gamma)$ of admissible functions is not emptv and*

$$k=\inf_{u\in\mathfrak{K}_\varphi^{(i)}}K(u).$$

Then k is a finite number, i.e. the functional $K(u)$ is bounded from below on $\mathfrak{K}_\varphi^{(i)}$ and there exists a unique function $u_0\in\mathfrak{K}_\varphi^{(i)}$ such that $K(u_0)=k$. In addition, any sequence $u_m\in\mathfrak{K}_\varphi^{(i)}$ for which $\lim_{m\to\infty}K(u_m)=k$ (such sequences are said to be minimizing) *satisfies the condition* $\lim u_m=u_0$ *in* $W_2^{(1)}$.[3]

Suppose now G^* is a domain such that $\overline{G}^*\subset G$, and suppose given on G a

(3)We note that variational problems in the case when the boundary conditions are given not on all of the boundary of the domain but only on part of it have previously been considered for domains with a sufficiently smooth boundary (see, for example, [10]).

continuously differentiable function that is equal to zero on $G\backslash G^*$. Such functions will be called *variations* and denoted by the symbols δu, δu_0, etc.

The function u_0 defined in Theorem [17.1] satisfies the *variational equation*

$$A(u_0, \delta u_0) - (f, \delta u_0) = 0. \tag{17.14}$$

For consider the function $\lambda(t) = K(u_0 + t\delta u_0)$. We have

$$\begin{aligned}\lambda(t) &= A(u_0 + t\delta u_0) - 2(f, u_0 + t\delta u_0)\\ &= t^2 A(\delta u_0) + 2[A(u_0, \delta u_0) - (f, \delta u_0)]t + A(u_0) - 2(f, u_0),\end{aligned}$$

and since $u_0 + t\delta u_0 \in \mathfrak{R}^{(i)}_{\varphi}$ for any real t, the function $\lambda(t)$ achieves a minimum at $t = 0$. This together with Fermat's theorem implies (17.14).

Thus we have completely studied the variational problem; namely, we have proved that in every nonempty class $\mathfrak{R}^{(i)}_{\varphi}$ it has a unique solution which minimizes the integral in question and satisfies the variational equation (17.14).

We proceed now to a study of the corresponding differential problem. To this end we impose additional conditions on the coefficients A_{ij}, B_i and C. We will assume that the functions A_{ij} and B_i together with all of their first order derivatives are bounded in absolute value on G.

In this case the *Euler equation* for $K(u)$ can be written in the form

$$\sum_{i=1}^{n}\sum_{j=1}^{n}\frac{\partial}{\partial x_j}\left(A_{ij}\frac{\partial u}{\partial x_i}\right) + qu + f = 0, \tag{17.15}$$

where

$$q = -C + \sum_{i=1}^{n}\frac{\partial B_i}{\partial x_i}. \tag{17.16}$$

For the sake of brevity we use the notation

$$L(u) \equiv \sum_{i=1}^{n}\sum_{j=1}^{n}\frac{\partial}{\partial x_j}\left(A_{ij}\frac{\partial u}{\partial x_i}\right) + qu. \tag{17.17}$$

Further, we will assume that a boundary function φ is given on all of the (accessible in the coordinate directions) boundary of G, and we will let simply $\mathfrak{R}_{\varphi}$ denote the set of all functions $u \in W_2^{(1)}$ such that $u|_{\Gamma} = \varphi$.

We note that all of the above assertions concerning a solution of the variational problem in the classes $\mathfrak{R}^{(i)}_{\varphi}$ obviously remain true for the class $\mathfrak{R}_{\varphi}$ (assuming that it is nonempty).

We now put

$$G_\eta = \{P: \rho(P, E^n \setminus G) > \eta\}.$$

Suppose

$$u \in W_2^{(1)}(G), \quad v \in W_2^{(1)}(G),$$

$$v \in W_2^{(2)}(G_\eta) \quad \text{for any} \quad \eta > 0 \text{ and } u|_\Gamma = 0, \tag{17.18}$$

or $u \in W_2^{(1)}(G)$, $v \in W_2^{(1)}(G)$ and

$$v|_\Gamma = 0, \left.\frac{\partial v}{\partial x_i}\right|_\Gamma = 0, \; i = 1, 2, \ldots, n. \tag{17.19}$$

Then the following formula of *Green's formula type* holds:

$$A(u, v) = -(u, L(v)). \tag{17.20}$$

To prove this formula under the boundary conditions (17.18) we first note that if a function $v \in W_2^{(2)}(G_\eta)$ for any $\eta > 0$, then it can for fixed i and k $(i, k = 1, \cdots, n)$ be modified on a set of measure zero so that for almost all $x \in \mathrm{pr}_i\, G$ the function $\partial v/\partial x_k$ will be absolutely continuous in x_i on each segment contained in G and lying on a straight line perpendicular to the hyperplane $x_i = 0$ and passing through one of the indicated points $x \in \mathrm{pr}_i\, G$.

In order to see this it suffices to partition G into a countable number of cubes Q, with edges parallel to the coordinate axes, such that $\bar{Q} \subset G$ and hence $v \in W_2^{(2)}(Q)$. One should then modify the function v on a set of measure zero in each cube so that the function $\partial v/\partial x_k$ is absolutely continuous in x_i on almost all segments perpendicular to the hyperplane $x_i = 0$ and contained in the cube $\bar{Q}$ in question. As a result, one obtains the required modification of v on a set of measure zero in G.

We proceed with the proof of (17.20). According to (17.3) we have

$$A(u, v) = \sum_{i=1}^{n} \sum_{k=1}^{n} \int A_{ik} \frac{\partial u}{\partial x_i} \frac{\partial u}{\partial x_k} dG + \sum_{i=1}^{n} \int B_i \frac{\partial u}{\partial x_i} v dG$$
$$+ \sum_{i=1}^{n} \int B_i \frac{\partial v}{\partial x_i} u dG + \int C u v dG. \tag{17.21}$$

Consider a term with the first sum in the right side with fixed indices i and k. We make the above indicated modification of v on a set of measure zero and we modify the function u so as to make it absolutely continuous in x_i on almost all segments perpendicular to the hyperplane $x_i = 0$ and contained in G.

It can be proved (see the author's paper [16]) that the domain G can be represented as the union of not more than a countable number of measurable sets E_m for which there exist constants $h_m > 0$ such that the intersection of each E_m with a straight line perpendicular to the hyperplane $x_i = 0$ is either the empty set or an interval of length greater than h_m the ends of which lie on the boundary of G. Thus $G = \bigcup_m E_m$. We fix one of the sets E_m.

Let $(a(x), b(x))$ be the interval of intersection of E_m with the straight line perpendicular to the hyperplane $x_i = 0$ and passing through a point $x \in \mathrm{pr}_i E_m$ (here $\mathrm{pr}_i E_m$ denotes the orthogonal projection of E_m onto the hyperplane $x_i = 0$). For any h and l such that $h + l < h_m/2$ let $E_m^{(h,l)}$ denote the set of all segments of the form $[a(x) + h, b(x) - l]$ for all possible $x \in \mathrm{pr}_i E_m$. It is obvious that for any sequences h_ν and l_μ, $\nu, \mu = 1, 2, \cdots$, such that $\lim_{\nu\to\infty} h_\nu = \lim_{\mu\to\infty} l_\mu = 0$ we have

$$\bigcup_{\nu,\mu=1}^{\infty} E_m^{(h_\nu, l_\mu)} = E_m .$$

We now consider the term $A_{ik}(\partial u/\partial x_i)(\partial v/\partial x_k)$ of $A(u, v)$ on $E_m^{(h,l)}$ and, using Fubini's theorem, integrate it by parts with respect to x_i over $[a(x) + h, b(x) - l]$, where x ranges over a set of total measure in $\mathrm{pr}_i E_m^{(h,l)} = \mathrm{pr}_k E_m$:

$$\int_{E_m^{(h,l)}} \!\!\cdots\!\int A_{ik} \frac{\partial u}{\partial x_i}\frac{\partial v}{\partial x_k}\, dx_1 \ldots dx_n = \int_{\mathrm{pr}_i E_m} \!\!\cdots\!\int \left[A_{ik}\frac{\partial v}{\partial x_k} u\right]_{x_i = a(x)+h}^{x_i = b(x)-l} dx_1 \ldots \widehat{dx_i} \ldots dx_n$$

$$- \int_{E_m^{(h,l)}} \!\!\cdots\!\int u \frac{\partial}{\partial x_i}\left(A_{ik}\frac{\partial v}{\partial x_k}\right) dx_1 \ldots dx_n, \quad x \in \mathrm{pr}_i E_m. \text{(4)} \qquad (17.22)$$

Let us estimate the first integral of the right side.(5) Suppose $|A_{ik}| \leqslant A$ on G, where A is a constant. We have, for example,

(4) The sign ^ over a symbol means that this symbol should be omitted.

(5) The method presented below for estimating this integral in the case of domains with a piecewise smooth boundary was first encountered by the author for the Dirichlet integral in the works of Nikol'skiĭ.

$$J(h) = \int_{\mathrm{pr}_l E_m}\!\!\!\ldots\!\int \left[A_{ik} u \frac{\partial v}{\partial x_k}\right]_{x_l = a(x)+h} dx_1 \ldots \hat{dx_l} \ldots dx_n$$

$$\leqslant A \sqrt{\int_{\mathrm{pr}_l E_m}\!\!\!\ldots\!\int \left[\left(\frac{\partial v}{\partial x_k}\right)^2\right]_{x_l = a(x)+h} dx_1 \ldots \hat{dx_l} \ldots dx_n}$$

$$\times \sqrt{\int_{\mathrm{pr}_l E_m}\!\!\!\ldots\!\int [u^2]_{x_l = a(x)+h}\, dx_1 \ldots \hat{dx_l} \ldots dx_n}.$$

Let

$$I_1(h) = \int_{\mathrm{pr}_l E_m}\!\!\!\ldots\!\int \left[\left(\frac{\partial v}{\partial x_k}\right)^2\right]_{x_l = a(x)+h} dx_1 \ldots \hat{dx_l} \ldots dx_n,$$

$$I_2(h) = \int_{\mathrm{pr}_l E_m}\!\!\!\ldots\!\int [u^2]_{x_l = a(x)+h}\, dx_1 \ldots \hat{dx_l} \ldots dx_n.$$

The function $I_1(h)$ is summable over $[0, h_m]$ since

$$\int_0^{h_m} I_1(h)\,dh = \int_{\mathrm{pr}_l E_m}\!\!\!\ldots\!\int \left\{\int_0^{h_m} \left[\left(\frac{\partial v}{\partial x_k}\right)^2\right]_{x_l = a(x)+h} dh\right\} dx_1 \ldots \hat{dx_l} \ldots dx_n$$

$$= \int_{\mathrm{pr}_l E_m}\!\!\!\ldots\!\int \int_{a(x)}^{a(x)+h_m} \left(\frac{\partial v}{\partial x_k}\right)^2 dx_1 \ldots dx_l \ldots dx_n \leqslant \int_{E_m}\!\!\ldots\!\int \left(\frac{\partial v}{\partial x_k}\right)^2 dx_1 \ldots dx_n$$

$$\leqslant D(v) < +\infty.$$

Hence by virtue of the convergence criterion for improper integrals of non-negative functions $(I_1(h) \geqslant 0)$ there exists a sequence h_ν, $\nu = 1, 2, \cdots$, such that $\lim_{\nu \to \infty} h_\nu = 0$ and $I_1(h_\nu) < 1/h_\nu$, i.e.

$$I_1(h_\nu) = O\left(\frac{1}{h_\nu}\right).$$

Further, from the condition $u|_\Gamma = 0$ it follows that

$$u|_{x_l = a(x)+h} = \int_{a(x)}^{a(x)+h} \frac{\partial u}{\partial x_l}\, dx_l$$

for almost all $x \in \mathrm{pr}_i\, E_m$. Applying Hölder's inequality, we get

$$I_2(h) = \int \underset{\mathrm{pr}_i E_m}{\cdots} \int \left(\int_{a(x)}^{a(x)+h} \frac{\partial u}{\partial x_i} dx_i \right)^2 dx_1 \dots \widehat{dx_i} \dots dx_n$$

$$\leqslant h \int \underset{\mathrm{pr}_i E_m}{\cdots} \int \int_{a(x)}^{a(x)+h} \left(\frac{\partial u}{\partial x_i} \right)^2 dx_1 \dots dx_i \dots dx_n$$

$$\leqslant h \int \underset{E_m \setminus E_m^{(h)}}{\cdots} \int \left(\frac{\partial u}{\partial x_i} \right)^2 dx_1 \dots dx_n = h\varepsilon(h),$$

where $\lim_{h \to 0} \epsilon(h) = 0$ since the integral

$$\int \underset{E_m}{\cdots} \int \left(\frac{\partial u}{\partial x_i} \right)^2 dx_1 \dots dx_n \leqslant D(u)$$

and is consequently finite. Thus $I_2(h) = o(h)$. Hence

$$J(h_\nu) \leqslant A \sqrt{I_1(h_\nu)} \sqrt{I_2(h_\nu)} = \sqrt{O\left(\frac{1}{h_\nu}\right) o(h_\nu)} = o(1).$$

The convergence to zero of the corresponding integral when $x_i = b(x) = l_\mu$ for a sequence l_μ, $\mu = 1, 2, \cdots$, such that $\lim_{\mu \to \infty} l_\mu = 0$ is proved completely analogously.

Passing to the limit now in equality (17.22) with respect to the sequences $a(x) + h_\nu$ and $b(x) - l_\mu$ ($\nu = 1, 2, \cdots ; \mu = 1, 2, \cdots$), we have

$$\int \underset{E_m}{\cdots} \int A_{ik} \frac{\partial u}{\partial x_i} \frac{\partial v}{\partial x_k} dx_1 \dots dx_n = - \int \underset{E_m}{\cdots} \int u \frac{\partial}{\partial x_i} \left(A_{ik} \frac{\partial v}{\partial x_k} \right) dx_1 \dots dx_n.$$

Finally, summing over $m = 1, 2, \cdots$, we arrive at the formula

$$\int \underset{G}{\cdots} \int A_{ik} \frac{\partial u}{\partial x_i} \frac{\partial v}{\partial x_k} dx_1 \dots dx_n = - \int \underset{G}{\cdots} \int u \frac{\partial}{\partial x_i} \left(A_{ik} \frac{\partial v}{\partial x_k} \right) dx_1 \dots dx_n. \quad (17.23)$$

The term $B_i \partial u / \partial x_i$ can be integrated by parts with respect to x_i directly over all of the segments $[a_{k_x}^{(i)}, b_{k_x}^{(i)}]$ for almost all $x \in \mathrm{pr}_i\, G$ and all $k_x = 1, 2, \cdots$. As a result, we get

$$\int_G \cdots \int B_i \frac{\partial u}{\partial x_i} v \, dx_1 \ldots dx_n = - \int_G \cdots \int u \frac{\partial}{\partial x_i} (B_i v) \, dx_1 \ldots dx_n. \tag{17.24}$$

Substituting (17.23) and (17.24) into (17.21), we obtain formula (17.20) under the boundary conditions (17.18).

In the case of boundary conditions (17.19) formula (17.20) can be proved by a direct integration by parts without additional estimates.

The following assertion is easily obtained with the use of formula (17.20): a function u satisfies the Euler equation (17.15) if it satisfies the variational equation

$$A(u, \delta u) - (f, \delta u) = 0 \tag{17.14}$$

or if it is a solution of the above-mentioned variational problem (see Theorem [17.1]) and belongs to $W_2^{(2)}(G_\eta)$ for any $\eta > 0$. In fact, from (17.14) and (17.20) we have

$$(L(u) + f, \delta u) = 0.$$

Hence by virtue of the arbitrariness of the variation δu it follows according to the fundamental lemma of the calculus of variations that $L(u) + f = 0$ almost everywhere in G.

Formula (17.20) also implies that a solution of the equation $Lu + f = 0$ in the class $\mathfrak{R}_\varphi$ is unique among the functions $u \in W_2^{(2)}(G_\eta)$ for any $\eta > 0$. For suppose u_1 and u_2 are two such solutions of $L(u) + f = 0$. Then

$$(u_1 - u_2)/_\Gamma = 0 \text{ and } (L(u_1 - u_2), \ u_1 - u_2) = 0.$$

and hence, by (17.20), $A(u_1 - u_2, u_1 - u_2) = 0$. But then inequality (17.4) implies $D(u_1 - u_2) = 0$. Therefore, on the basis of (17.10), we get $\|u_1 - u_2\|_{L_2} = 0$. From this result we get that $u_1 = u_2$ almost everywhere in G.

In order to simplify the presentation and because of our subsequent goals we confine ourselves to a consideration of the differential properties of the solution of the variational problem for a somewhat more special case. In fact, we will assume in addition that $A_{ii} = 1$ and $A_{ij} = 0$ for $i \neq j$, $i.\ i = 1, \cdots, n$. Then

$$L(u) = \Delta u + qu. \tag{17.25}$$

We will also assume that the functions B_i $(i = 1, \cdots, n)$ have bounded derivatives up to second order inclusively, that the function C together with its derivatives of first order is bounded and hence that the function q together with its derivatives of first order is bounded and measurable. Finally, we will assume that the function f has derivatives belonging to L_2.

It should be noted that the twice differentiability property of the solution u_0 of the variational problem is an internal local property. Therefore the proof of differentiability in our case of an arbitrary bounded domain G is not essentially different from similar proofs in cases considered earlier by Sobolev [2], Kondrašov [10] and Višik [42]. For the sake of completeness of the presentation we will nevertheless give a proof of the twice differentiability of u_0 since it is sufficiently elementary. In this connection we will only make use of some properties of the volume potential in n-dimensional space. A direct proof of these properties can be found in Sobolev [2].

We require the following lemma on integrals of potential type, which in essence was also proved earlier by Sobolev.

Let $\psi(t)$ denote a monotone decreasing function that is infinitely differentiable on the real line and such that $\psi(t) = 1$ for $t \leqslant ½$ and $\psi(t) = 0$ for $t \geqslant 1$. Further, suppose $\eta > 0$ and let, as usual,

$$G_\eta = \{P : \rho(P, E^n \setminus G) > \eta\},$$

$$Q = (\xi_1, \ldots, \xi_n) \in G, \quad P = (x_1, \ldots, x_n) \in G_\eta, \quad r = \rho(P, Q).$$

[17.2.1]. *Suppose* $1 \leqslant p \leqslant \infty$ *and*

$$\Psi_h(P) = \int \cdots \int_{r \leqslant h} \frac{1}{r^{n-2}} \psi\left(\frac{r}{h}\right) F(Q) dv_Q, \tag{17.26}$$

where $P \in G_\eta$ *and* $0 < h < \eta$. *Then the following assertions hold.*

1°. *If* $F \in L_p(G)$, *then* $\Psi_h \in L_{p'}(G_\eta)$, *where* $p' = \infty$ *if* $p > n/2$ *and* $p' < p^* = np/(n-2p)$ *if* $p \leqslant n/2$;(6) *here* $\lim_{h\to\infty} \|\Psi_h\|_{L_{p'}(G_\eta)} = 0$.

2°. *If* $F \in L_2(G)$, *then* Ψ_h *has weak derivatives with respect to all of its arguments, with*

$$\frac{\partial \Psi_h}{\partial x_i} = \int \cdots \int_{r \leqslant h} \frac{\partial}{\partial x_i}\left[\frac{1}{r^{n-2}} \psi\left(\frac{r}{h}\right)\right] F(Q)\, dv_Q. \tag{17.27}$$

If, in addition, F *has a weak derivative* $\partial F/\partial x_i \in L_2$, *then*

$$\frac{\partial \Psi_h}{\partial x_i} = \int \cdots \int_{r \leqslant h} \frac{1}{r^{n-2}} \psi\left(\frac{r}{h}\right) \frac{\partial F}{\partial \xi_i}\, dv_Q. \tag{17.28}$$

---

(6) In this case there always exists a $p' > p$, since $p^* - p = p^2/(n/2 - p) > 0$.

PROOF. The assertion $\Psi_h \in L_p'(G_\eta)$ is proved in [2] (Theorem 2, pages 43–45), the condition of uniform convergence to zero of the norm $\|\Psi_h\|_{L_{p'}(G_\eta)}$ following from the estimates appearing there. The same theorem implies the existence of the integral on the right side of equality (17.27) and even the membership of the function defined by it in a space $L_p(G_\eta)$. Let us show that this integral actually is the derivative $\partial\Psi_h/\partial x_i$. Suppose $\omega = \omega(P)$ is a continuously differentiable function on G that is identically equal to zero outside a closed domain lying in G_η. Then

$$\int \cdots \int_{r \leqslant h} \left[\omega \frac{\partial}{\partial x_i}\left(\psi\left(\frac{r}{h}\right)\frac{1}{r^{n-2}}\right) + \psi\left(\frac{r}{h}\right)\frac{1}{r^{n-2}}\frac{\partial \omega}{\partial x_i}\right] dx_1 \ldots dx_n = 0. \quad (17.29)$$

For suppose $\epsilon > 0$. Applying Green's theorem, we get

$$\int \cdots \int_{\epsilon \leqslant r \leqslant h} \left[\omega \frac{\partial}{\partial x_i}\left(\psi\left(\frac{r}{h}\right)\frac{1}{r^{n-2}}\right) + \psi\left(\frac{r}{h}\right)\frac{1}{r^{n-2}}\frac{\partial \omega}{\partial x_i}\right] dv_Q$$

$$= -\int \cdots \int_{r=\epsilon} \frac{\cos(\widehat{nx_i})}{r^{n-2}}\psi\left(\frac{r}{h}\right)\omega\, ds \to 0 \text{ for } \epsilon \to 0.$$

Therefore

$$\int \cdots \int_{G_\eta} \left(\omega \frac{\partial \Psi_h}{\partial x_i} + \Psi_h \frac{\partial \omega}{\partial x_i}\right) dv_P = \int \cdots \int_G F(Q) \left\{\int \cdots \int_{r \leqslant h} \left[\omega \frac{\partial}{\partial x_i}\left(\psi\left(\frac{r}{h}\right)\frac{1}{r^{n-2}}\right)\right.\right.$$

$$\left.\left. + \psi\left(\frac{r}{h}\right)\frac{1}{r^{n-2}}\frac{\partial \omega}{\partial x_i}\right] dv_P\right\} dv_Q = 0.$$

But this means that the function $\partial\Psi_h/\partial x_i$ defined by (17.27) is a weak derivative of Ψ_h. Further, noting that the function $\psi(r/h)/r^{n-2}$ has a weak derivative by virtue of (17.2), we get

$$\int \cdots \int_{r \leqslant h} \frac{\partial}{\partial x_i}\left[\frac{1}{r^{n-2}}\psi\left(\frac{r}{h}\right)\right] F(Q)\, dv_Q$$

$$= -\int \cdots \int_{r \leqslant h} \frac{\partial}{\partial \xi_i}\left[\frac{1}{r^{n-2}}\psi\left(\frac{r}{h}\right)\right] F(Q)\, dv_Q = \int \cdots \int_{r \leqslant h} \frac{1}{r^{n-2}}\psi\left(\frac{r}{h}\right)\frac{\partial F}{\partial \xi_i} dv_Q$$

$$-\int \cdots \int_{r \leqslant h} \frac{\partial}{\partial \xi_i}\left[\frac{1}{r^{n-2}}\psi\left(\frac{r}{h}\right) F\right] dv_Q = \int \cdots \int_{r \leqslant h} \frac{1}{r^{n-2}}\psi\left(\frac{r}{h}\right)\frac{\partial F}{\partial \xi_i} dv_Q,$$

since the second integral is equal to zero by the definition of weak derivatives, i.e. we have obtained (17.28). The lemma is proved.

We return to the proof of the differentiability of the solution u_0 of the variational problem. Consider the variation defined by the equality

$$\delta u_0 = \frac{1}{r^{n-2}}\left[\psi\left(\frac{r}{h_1}\right) - \psi\left(\frac{r}{h_2}\right)\right], \quad 0 < h_1 < h_2 < \eta \text{ (7)}$$

(in which $1/r^{n-2}$ is replaced by $\ln(1/r)$ if $n = 2$). Clearly, δu_0 depends on the point Q.

Let us fix the point Q. Then $\delta u_0 = 0$ for $r < h_1/2$ and therefore δu_0 together with all of its derivatives is continuous on G. In addition, $\delta u_0 = 0$ on $G \setminus G_{\eta - h_2}$. As a result, formula (17.14) can be applied:

$$A(u_0, \delta u_0) - (f, \delta u_0) = 0.$$

Then according to (17.20)

$$(u_0, L(\delta u_0)) + (f, \delta u_0) = 0$$

or, more explicitly,

$$\underset{r \leqslant h_1}{\int \cdots \int} \Delta\left[\frac{1}{r^{n-2}}\psi\left(\frac{r}{h_1}\right)\right] u_0(Q)\, dv_Q - \underset{r \leqslant h_1}{\int \cdots \int} \frac{1}{r^{n-2}}\psi\left(\frac{r}{h_1}\right)(f + qu_0)\, dv_Q$$
$$= \underset{r \leqslant h_2}{\int \cdots \int} \Delta\left[\frac{1}{r^{n-2}}\psi\left(\frac{r}{h_2}\right)\right] u_0(Q)\, dv_Q - \underset{r \leqslant h_2}{\int \cdots \int} \frac{1}{r^{n-2}}\psi\left(\frac{r}{h_2}\right)(f + qu_0)\, dv_Q. \tag{17.30}$$

The function

$$\Delta\left[\frac{1}{r^{n-2}}\psi\left(\frac{r}{h_1}\right)\right]$$

is infinitely continuously differentiable, and, as is shown in [2] (page 93),

$$\frac{1}{(n-2)\, s_n}\int \Delta\left[\frac{1}{r^{n-2}}\psi\left(\frac{r}{h}\right)\right] dv_Q = 1, \tag{17.31}$$

---

(7) In the case of the general selfadjoint elliptic equation the variation δu is constructed analogously, starting from the singularities of the corresponding elementary solutions of the operator $L(u)$ (in the present case we have simply taken r^{2-n} as the elementary solution of Laplace's equation). The rest of the proof of the twice differentiability of u_0 proceeds in completely the same way as in the present case.

where $s_n = 2\pi^{n/2}/\Gamma(n/2)$ is the surface of the unit ball in n-dimensional space. Therefore the first integrals in each side of (17.30) are averagings, while the second integrals are of potential type, and since according to our assumptions $\psi(r/h_i)(f - qu_0) \in L_2$ $(i = 1, 2)$, these integrals are finite in every case and hence the separation into the written summands is legitimate.

Formula (17.30) shows that the expression in each side of this equality does not depend on h.

Let

$$\Phi_h(P) = \frac{1}{(n-2)s_n} \int \cdots \int_{r \leqslant h} \Delta \left[\frac{1}{r^{n-2}} \psi\left(\frac{r}{h}\right) \right] u_0(Q)\, dv_Q,$$

$$\Psi_h(P) = -\frac{1}{(n-2)s_n} \int \cdots \int_{r \leqslant h} \frac{1}{r^{n-2}} \psi\left(\frac{r}{h}\right) (f + qu_0)\, dv_Q.$$

Further, let $u(P) = \Phi_h(P) + \Psi_h(P)$.

This equality uniquely defines the function $u(P)$ at each point $P \in G$ for sufficiently small h.

Let us show that $u = u_0$ almost everywhere on G.(8)

By virtue of the fact that $\Phi_h(P)$ is an averaging of u_0 we have for any $\eta > 0$

$$\lim_{h \to 0} \| \Phi_h - u_0 \|_{L_2(G_\eta)} = 0.$$

But according to Lemma [17.2.1] we also have for any $\eta > 0$

(8) It is natural to expect this since the following formula holds for any twice continuously differentiable function v on G:

$$v(P) = \frac{1}{(n-2)s_n} \int \cdots \int_{r \leqslant h} \Delta \left[\frac{1}{r^{n-2}} \psi\left(\frac{r}{h}\right) \right] v(Q)\, dv_Q + \frac{1}{(n-2)s_n} \int \cdots \int_{r \leqslant h} \frac{1}{r^{n-2}} \psi\left(\frac{2}{h}\right) \Delta v\, dv_Q.$$

It is obtained in the usual way by applying Green's formula

$$\int (u\Delta v - v\Delta u)\, dG = \int \left(u \frac{\partial v}{\partial n} - v \frac{\partial u}{\partial n} \right) ds$$

for the case $u = r^{2-n}\psi(2/h)$. In our case $\Delta u_0 = -f - qu_0$.

Hence

$$\lim_{h\to 0} \|(\Phi_h + \Psi_h) - u_0\|_{L_2(G_\eta)} = 0.$$

But, as we have seen, the sum $\Phi_h + \Psi_h$ does not depend on h, and hence the latter relation can hold only if $u_0 = \Phi_h + \Psi_h$ almost everywhere on G for any $\eta > 0$. It follows that

$$u_0 = \Phi_h + \Psi_h \tag{17.32}$$

almost everywhere on G.

As already noted above, the function Φ_h is an averaging in the sense introduced in Sobolev's monograph [2], and therefore has derivatives of all orders with respect to all of its arguments.

The proof of the existence of the derivatives of Ψ_h is based on Lemma [17.2.1]. Noting that $f - qu_0$ has a first order derivative belonging to L_2, we see that

$$\frac{\partial \Psi_h}{\partial x_i} = \frac{1}{(n-2)\, s_n} \int \cdots \int_{r \leq h} \frac{1}{r^{n-2}} \psi\left(\frac{r}{h}\right) \frac{\partial}{\partial \xi_i} (f - qu_0)\, dv_Q \tag{17.33}$$

exists and in every case $\partial \Psi_h / \partial x_i \in L_2(G_\eta)$.(9) Therefore by virtue of (17.33) there exists $\partial u / \partial x_i \in L_2(G_\eta)$, and hence $\partial (f - qu_0)/\partial \xi_i \in L_2(G_\eta)$ $(i = 1, \cdots, n)$. But then the same lemma and (17.29) imply that there exists $\partial^2 \Psi_h / \partial x_i^2 \in L_2(G_\eta)$, and hence $\partial^2 u_0 / \partial x_i^2 \in L_2(G_\eta)$, $\eta > 0$ $(i = 1, \cdots, n)$.

From this result it follows, as we have already seen earlier, that $L(u_0) + f = 0$ almost everywhere on G.

We have thus proved the following theorem.

[17.2]. *Suppose a function u_0 effects a minimum of the functional*

$$K(u) = \int \left[\sum_{i=1}^{n} \left(\frac{\partial u}{\partial x_i}\right)^2 + 2 \sum_{i=1}^{n} B_i \frac{\partial u}{\partial x_i} u + Cu^2 - 2fu \right] dG \tag{17.34}$$

in a function class $\mathfrak{K}_\varphi$, the functions B_i $(i = 1, \cdots, n)$ having bounded derivatives up to second order inclusively, the function C together with its first order derivatives being bounded and the function f having derivatives up to second order inclusively

(9) Indeed, $\partial \Psi_h / \partial x_i \in L_p(G)$ for some $p > 2$ (see [2]).

belonging to L_2.(10) *Then the function* $u_0 \in \mathfrak{K}_\varphi$ *has derivatives of second order*(11) *belonging to* $L_2(G_\eta)$ *for any* $\eta > 0$ *and satisfies almost everywhere on* G *the Euler equation for* $K(u)$:

$$\Delta u + qu = -f, \text{ where } q = -C + \sum_{i=1}^{n} \frac{\partial B_i}{\partial x_i}.$$

A solution of this equation in a class $\mathfrak{K}_\varphi$ *is unique among the functions of* $\mathfrak{K}_\varphi$ *belonging to* $W_2^{(2)}(G_\eta)$ *for any* $\eta > 0$.

REMARK. If it is assumed in addition that f has derivatives of second order belonging to L_p, where $p > n/2$, and q has bounded derivatives of second order, the function u_0 is twice continuously differentiable and hence is a solution of the Euler equation in the classical sense. This is proved by a method of Kondrašov. It consists in the following.

We already know that in this case

$$\frac{\partial^2 \Psi_h}{\partial x_i^2} = -\frac{1}{(n-2) s_n} \int \cdots \int_{h \leq r} \frac{1}{r^{n-2}} \psi\left(\frac{r}{h}\right) \frac{\partial^2}{\partial \xi_i^2}(f + qu_0)\, dv_Q,$$

which implies according to assertion 1° of Lemma [17.2.1] and formula (17.32) that $\partial^2 u_0/\partial x_i^2 \in L_{p_1}$, where, for example, $p_1 = p + p^2/(n - 2p)$. If $p_1 > n/2$, the same lemma implies at once that $\partial^2 \Psi_h/\partial x_i^2 \in L_\infty(G_{\eta_1})$ and hence that $\partial^2 u_0/\partial x_i^2 \in L_\infty(G_{\eta_1})$ for any $\eta_1 > 0$ $(i = 1, \cdots, n)$. If on the other hand $p_1 \leq n/2$, the lemma together with the fact that $\partial^2(f - qu_0)/\partial \xi_i^2 \in L_{p_1}(G_{\eta_1})$ implies that $\partial^2 \Psi_h/\partial x_i^2 \in L_{p_2}(G_{\eta_2})$, where $\eta_2 > \eta_1$ and, for example,

$$p_2 = p_1 + \frac{p_1^2}{2\left(\frac{n}{2} - p_1\right)} \geq p + \frac{p^2}{\left(\frac{n}{2} - p\right)}.$$

Thus. after k steps, we get

$$p_k \geq p + \frac{kp^2}{2\left(\frac{n}{2} - p\right)}.$$

---

(10) The assertion of the theorem will be preserved in the case of the general function (17.11) if one requires in addition that the coefficients A_{ik} be twice continuously differentiable in the closed domain $\overline{G}$.

(11) All of the derivatives are understood in the weak sense.

Choosing k so that $p_k > n/2$, we get that $\partial^2\Psi_h/\partial x_i^2 \in L_{p_k}(G_{\eta_k})$ for some $\eta_k > \eta_{k-1} > \cdots > \eta_1 > 0$. But then $\partial^2 u_0/\partial x_i^2 \in L_{p_k}(G_{\eta_k})$, which implies by the same lemma again that $\partial^2\Psi_h/\partial x_i^2 \in L_\infty(G_{\eta_k})$ and hence that $\partial^2 u_0/\partial x_i^2 \in L_\infty(G_{\eta_k})$ ($i = 1, \cdots, n$), the number $\eta_k > 0$ being arbitrary by virtue of the arbitrariness of $\eta_1 > 0$.

It can easily be seen by the same method that in the case $q \equiv 0$ there are naturally even fewer restrictions that need be imposed on f to obtain a classical solution.

If the coefficients q and f are analytic it can be shown in a similar way that u_0 has continuous derivatives of arbitrarily high order and hence, as is well known (see [39, 40]), is analytic.

By using the more exact results of Calderon and Zygmund [45] on integrals of potential type, one can lower the order of the derivatives of the coefficients of the equation required in the conditions of Theorem [17.2].

We note that the methods developed in the present section can be carried over in a natural way to the case of elliptic equations of higher order, for example, to a polyharmonic equation.

In conclusion we make some remarks concerning the nonemptiness of the class $\mathfrak{K}_\varphi$. This problem has already been studied to a sufficient extent in the literature, and necessary and sufficient conditions in the presence of a known smoothness of the boundary have been obtained for it.

A necessary condition on the boundary function consists in the following (see Sobolev [2]).

If a point P of the boundary Γ of the domain G is proper(12) and if $u \in W_2^{(1)}$, there exists a neighborhood O of P in Γ such that $u|_O \in H_{2(n-1)}^{(1/2)}(O)$.

A sufficient condition for the nonemptiness of $\mathfrak{K}_\varphi$ consists in the following (see Nikol'skiĭ [4]).

If the boundary Γ of the domain G consists of a finite number of manifolds of smoothness $k \geqslant 2$ and $\varphi \in H_{2(n-1)}^{(1/2+\epsilon)}(\Gamma)$, $\epsilon > 0$, then the class $\mathfrak{K}_\varphi$ is nonempty.

We note that in a number of cases it is possible, by using another terminology, to formulate the indicated necessary conditions in such a way that they are at the same time also sufficient (see, for example, [46, 47, 11, 48]).

---

(12) A point P of the boundary Γ of a domain G is said to be proper if there exists a neighborhood of this point in Γ which is an $(n-1)$-dimensional manifold of smoothness $k \geqslant 2$.

§18. General plan for the solution of degenerate equations by the variational method

We introduce some notation which will be employed from now on in the present chapter. G is a domain of the space E^n, K is its boundary and $\sigma = \sigma(x_1, \cdots, x_n)$ is a thrice continuously differentiable positive function on G. For any pair of functions u_1 and u_2 defined on G together with their weak derivatives of first order we put

$$D_\alpha(u_1, u_2) = \int_G \cdots \int \sigma^\alpha \sum_{i=1}^{n} \frac{\partial u_1}{\partial x_i} \frac{\partial u_2}{\partial x_i} dx_1 \ldots dx_n, \tag{18.1}$$

and, in particular,

$$D_\alpha(u,u) = D_\alpha(u). \tag{18.2}$$

For a twice weakly differentiable function u on G we set

$$L_\alpha(u) = \sum_{i=1}^{n} \frac{\partial}{\partial x_i}\left(\sigma^\alpha \frac{\partial u}{\partial x_i}\right). \tag{18.3}$$

Clearly,

$$L_\alpha(u) = 0 \tag{18.4}$$

is the Euler equation for the functional $D_\alpha(u)$.

The remainder of the present chapter is devoted to a study of the equations of form (18.3). The number α will be called the degeneracy exponent. As noted in the Introduction, when $n = 2$, for example, some equations of the type

$$y^\beta \frac{\partial^2 z}{\partial x^2} + \frac{\partial^2 z}{\partial y^2} = 0,$$

reduce to an equation of form (18.3) when $\sigma = y$.

We study necessary and sufficient conditions for the solution of the degenerate equations (18.4) by the variational method. The sufficient conditions are connected with questions of nonemptiness of the corresponding classes of admissible functions. Therefore the proofs given below are based on the methods of extending functions developed in the preceding sections for the case of sufficiently smooth boundaries. In connection with this we will always assume in the sequel a known smoothness of the boundary of the domain, which permits us to speak of the boundary values of a function not only in the sense of convergence almost everywhere, as was done in §17, but also in the sense of convergence in the mean. It should be noted, however, that questions analogous to the questions considered in § 17 can be discussed in the case of degenerate equations for a wider class of domains than is done below if one confines oneself to convergence almost everywhere to the boundary values. This will not be done in order to provide a uniform presentation of all of the questions which will be touched upon.

For a function u defined on a domain G and a function φ defined on the boundary Γ we will write $u|_\Gamma = \varphi$ if for any point $P \in \Gamma$ there exists a neighborhood $U_\Gamma(P)$ of it such that

$$u|_{U_\Gamma(P)} = \varphi$$

in the sense of convergence in the mean.

A function u will be said to be $D_{\alpha,\varphi}$ admissible if

1°. $$u|_\Gamma = \varphi, \tag{18.5}$$

2°. $$D_\alpha(u) < \infty. \tag{18.6}$$

The set of all $D_{\alpha,\varphi}$ admissible functions is denoted by $D_{\alpha,\varphi}(G)$, while the set of all functions having weak derivatives on G and such that $D_\alpha(u) < \infty$ is denoted by $D_\alpha(u)$. The functions of $D_\alpha(G)$ will simply be said to be admissible.

Clearly, $D_\alpha(u_1, u_2) < \infty$ if $u_1 \in D_\alpha(G)$ and $u_2 \in D_\alpha(G)$.

For the sake of simplicity we use the notation $\bar\sigma = \sigma^\alpha$. A function v defined on G is said to be $\bar D_{\alpha,\varphi}$ admissible if $u = v\bar\sigma^{-1/2} \in D_{\alpha,\varphi}(G)$, while the set of all $\bar D_{\alpha,\varphi}$ admissible functions is denoted by $\bar D_{\alpha,\varphi}(G)$.

Let

$$\bar D_\alpha(v_1, v_2) = D_\alpha\left(\bar\sigma^{-\frac{1}{2}} v_1, \bar\sigma^{-\frac{1}{2}} v_2\right), \tag{18.7}$$

and, in particular,

$$\bar D_\alpha(v) = \bar D_\alpha(v, v). \tag{18.8}$$

We have

$$\bar D_\alpha(v) = \int_G \cdots \int \sum_{i=1}^n \left[\left(\frac{dv}{dx_i}\right)^2 - \frac{1}{\bar\sigma}\frac{\partial\bar\sigma}{\partial x_i}\frac{dv}{dx_i} v + \frac{1}{4\bar\sigma^2}\left(\frac{\partial\bar\sigma}{\partial x_i}\right)^2 v^2\right] dx_1 \ldots dx_n. \tag{18.9}$$

The Euler equation for $\bar D_\alpha(v)$ has the form

$$L_\alpha(v) = \Delta v + v \sum_{i=1}^{n} \frac{\bar\sigma_{x_i}^2 - 2\bar\sigma\,\bar\sigma_{x_i x_i}}{4\bar\sigma^2}, \tag{18.10}$$

and, in this connection,

$$\bar\sigma^{-\frac{1}{2}} L_\alpha(v) = L_\alpha\left(\bar\sigma^{-\frac{1}{2}} v\right). \tag{18.11}$$

All of this can be verified by a direct calculation.

By using the connection between the functionals $D_\alpha(u)$ and $\bar D_\alpha(v)$, we can easily extend the variational method to the case of the considered degenerate equation $L_\alpha(u) = 0$.

To this end we first note that a function $u \in D_\alpha(G)$ satisfies the inequalities

$$D_\alpha(u) \geqslant 0, \tag{18.12}$$

$$\frac{1}{\sqrt{n}} \sum_{i=1}^{n} \left\| \frac{\partial u}{\partial x_i} \right\|_{2(\sigma,\alpha)}^{(n)} \leqslant D_\alpha^{\frac{1}{2}}(u) \leqslant \sum_{i=1}^{n} \left\| \frac{\partial u}{\partial x_i} \right\|_{2(\sigma,\alpha)}^{n}. \tag{18.13}$$

The left inequality of (18.13) is obtained directly from the weighted Hölder inequality, while the right is obtained from the inequality $(\Sigma a_i)^{1/2} \leqslant \Sigma a_i^{1/2}$, $a_i > 0$. And if $u \in D_\alpha(G)$ and $v \in D_\alpha(G)$, we have

$$\left| D_\alpha^{\frac{1}{2}}(u) - D_\alpha^{\frac{1}{2}}(v) \right| \leqslant D_\alpha^{\frac{1}{2}}(u - v). \tag{18.14}$$

This follows from the fact that $D_\alpha^{1/2}(u)$ satisfies the triangle inequality, since $D_\alpha(u, v)$ is a symmetric nonnegative bilinear functional.

As for an inequality in the degenerate case that is analogous to (17.10), our proof of one below will be carried out separately for each considered type of domain and will depend on the character of the degeneracy of the equation. In the present section, however, we only prove propositions that are general for the variational method (i.e. are independent of the form of the domain and the character of the degeneracy). Thus the variational method as a whole receives its completion for degenerate equations only in the following sections.

[18.1]. *Suppose the class* $D_{\alpha,\varphi}(G)$ *is nonempty. Let* $d = \inf_{u \in D_{\alpha,\varphi}(G)} D_\alpha(u)$. *Then* d *is a finite nonnegative number. Suppose, further,*

$$u_m \in D_{\alpha,\varphi}(G) \text{ and } \lim_{m \to \infty} D_\alpha(u_m) = d.$$

Then $D_\alpha(u_{m+\mu} - u_m) \to 0$ *for* $m \to \infty$ *and* $\mu = 1, 2, \cdots$.

COROLLARY.

$$\left\| \frac{\partial u_{m+\mu}}{\partial x_i} - \frac{\partial u_m}{\partial x_i} \right\|^{(n)}_{2(\sigma,\alpha)} \to 0 \quad \text{for } m \to \infty, \ \mu = 1,2,\ldots \tag{18.15}$$

The proof is completely analogous to the proof of the analogous assertions of §17 (see the proof of Theorem [17.1]) if one notes that the identity

$$D_\alpha\left(\frac{u-v}{2}\right) = \frac{1}{2} D_\alpha(u) + \frac{1}{2} D_\alpha(v) - D_\alpha\left(\frac{u+v}{2}\right)$$

holds for any $u \in D_\alpha(G)$ and $v \in D_\alpha(G)$ and uses inequality (18.13), which is analogous to inequality (17.2).

[18.2]. *Suppose* $\lim_{m\to\infty} D_\alpha(u_m) = d$, $u_m \in D_{\alpha,\varphi}(G)$, $m = 1, 2, \cdots$ *(such sequences are said to be* minimizing) *and there exists a function* $u_0 \in D_{\alpha,\varphi}(G)$ *such that*

$$\lim_{m\to\infty} D_\alpha(u_0 - u_m) = 0.$$

Then $D_\alpha(u_0) = d$.

PROOF. We have

$$|D_\alpha(u_0) - D_\alpha(u_m)| \leqslant |D_\alpha^{\frac{1}{2}}(u_0) - D_\alpha^{\frac{1}{2}}(u_m)|\,|D_\alpha^{\frac{1}{2}}(u_0) + D_\alpha^{\frac{1}{2}}(u_m)|$$
$$\leqslant D_\alpha^{\frac{1}{2}}(u_0 - u_m)\left[D_\alpha^{\frac{1}{2}}(u_0) + D_\alpha^{\frac{1}{2}}(u_m)\right].$$

The theorem follows from this inequality by virtue of the boundedness from above of $D_\alpha(u_m)$.

In the sequel u_0 will always denote a function belonging to a fixed class $D_{\alpha,\varphi}(G)$ and such that $D_\alpha(u_0) = d$ (see [18.1]). We note that if a function u_0 effects a minimum of $D_\alpha(u)$ in the class of $D_{\alpha,\varphi}$ admissible functions, then the function

$$v_0 = \sigma^{-\frac{1}{2}} u_0 \tag{18.16}$$

effects a minimum of $D_\alpha(v)$ in the class of $D_{\alpha,\varphi}$ admissible functions (see (18.7)).

The question of the existence of a function u_0, as well as the question of non-emptiness of the class of $D_{\alpha,\varphi}$ admissible functions for a given boundary function φ, will be considered in the following sections for each concrete case separately. Here we prove only

[18.3]. *If* $D_\alpha(u_0) = D_\alpha(u) = d$, $u_0 \in D_{\alpha,\varphi}(G)$ *and* $u \in D_{\alpha,\varphi}(G)$, *then*

$$D_\alpha(u_0 - u) = 0.$$

PROOF. From the fact that $D_\alpha((u+u_0)/2) \geqslant d$ (since $(u_0+u)/2 \in D_{\alpha,\varphi}(G)$) we have

$$0 \leqslant D_\alpha\left(\frac{u_0-u}{2}\right) = \frac{1}{2} D_\alpha(u_0) + \frac{1}{2} D_\alpha(u) - D_\alpha\left(\frac{u_0+u}{2}\right) \leqslant \frac{d}{2} + \frac{d}{2} - d = 0.$$

Hence $D_\alpha(u_0 - u) = 0$.

It will be shown in the sequel that [18.3] implies the uniqueness of the solution of the variational problem of finding a function that effects a minimum of $D_\alpha(u)$ in the class $D_{\alpha,\varphi}(G)$.

[18.4]. *Suppose* δv_0 *is a twice continuously differentiable variation* (*i.e.* $\delta v_0 = 0$ *on* $G \setminus G_\eta$ *for some* $\eta > 0$).(13) *Then*

$$D_\alpha(v_0, \delta v_0) = 0.$$

COROLLARY. *If* δv_0 *is twice differentiable*,

$$\int_G \cdots \int v_0 L_\alpha(\delta v_0)\, dx_1 \ldots dx_n = 0.$$

PROOF. Consider the function

$$\lambda(t) = D_\alpha(v_0 + t\delta v_0) = D_\alpha(v_0) + 2tD_\alpha(v_0, \delta v_0) + t^2 D_\alpha(\delta v_0).$$

Clearly, $D_\alpha(v_0 + t\delta v_0) \geqslant d$ since $v_0 + t\delta v_0 \in D_{\alpha,\varphi}$, and hence $\lambda(t)$ has a minimum at $t = 0$. It follows that $D_\alpha(v_0, \delta v_0) = 0$. The Corollary is obtained directly from this equality by virtue of the definition of weak derivatives and the properties of the function δv_0 (by integrating by parts).

Suppose now $\psi(t)$ is the function introduced in Lemma [17.2.1].

We choose as the function δv_0 in [18.4] the function

$$\delta v_0 = \frac{1}{r^{n-2}}\left[\psi\left(\frac{r}{h_1}\right) - \psi\left(\frac{r}{h_2}\right)\right], \quad 0 < h_1 < h_2 < \eta.$$

Then, taking into account (18.10), we get from the Corollary to [18.4] that

(13) As always, $G_\eta = \{P: \rho(P, E^n \setminus G) > \eta\}$.

$$\int_{r \leq h_1} \cdots \int \Delta \left[\frac{1}{r^{n-2}} \psi\left(\frac{r}{h_1}\right)\right] v_0(\xi_1, \ldots, \xi_n)\, d\xi_1 \ldots d\xi_n$$

$$+ \sum_{i=1}^{n} \int_{r \leq h_1} \cdots \int \frac{1}{r^{n-2}} \psi\left(\frac{r}{h_1}\right) \frac{\bar{\sigma}_{\xi_i}^2 - 2\bar{\sigma}\, \bar{\sigma}_{\xi_i \xi_i}}{4\bar{\sigma}^2} v_0(\xi_1, \ldots, \xi_n)\, d\xi_1 \ldots d\xi_n$$

$$= \int_{r \leq h_2} \cdots \int \Delta \left[\frac{1}{r^{n-2}} \psi\left(\frac{r}{h_2}\right)\right] v_0(\xi_1, \ldots, \xi_n)\, d\xi_1 \ldots d\xi_n$$

$$+ \sum_{i=1}^{n} \int_{r \leq h_2} \cdots \int \frac{1}{r^{n-2}} \psi\left(\frac{r}{h_2}\right) \frac{\bar{\sigma}_{\xi_i}^2 - 2\bar{\sigma}\, \bar{\sigma}_{\xi_i \xi_i}}{4\bar{\sigma}^2} v_0(\xi_1, \ldots, \xi_n)\, d\xi_1 \ldots d\xi_n.$$

We see that neither side of this equality depends on h, and thus we arrive at a definition on G of the function

$$v(x_1, \ldots, x_n) = \frac{1}{(n-2) s_n} \int_{r < h} \cdots \int \Delta \left[\frac{1}{r^{n-2}} \psi\left(\frac{r}{h}\right)\right] v_0(\xi_1, \ldots, \xi_n)\, d\xi_1 \ldots d\xi_n$$

$$+ \frac{1}{(n-2) s_n} \sum_{i=1}^{n} \int_{r < h} \cdots \int \frac{1}{r^{n-2}} \psi\left(\frac{r}{h}\right) \frac{\bar{\sigma}_{\xi_i}^2 - 2\bar{\sigma}\, \bar{\sigma}_{\xi_i \xi_i}}{4\bar{\sigma}^2} v_0(\xi_1, \ldots, \xi_n)\, d\xi_1 \ldots d\xi_n, \tag{18.17}$$

which is uniquely defined at each point of G for sufficiently small $h > 0$. We note that according to formula (18.10), if v_0 were a classical solution of the differential equation $L_\alpha(v) = 0$, it would be exactly equal to the left side of (18.17). In the general case the following theorem holds.

[18.5]. *Suppose* v_0 *and* v *are the functions defined by* (18.16) *and* (18.17) *respectively. Then* $v = v_0$ *almost everywhere on* G.

PROOF. It suffices to show that $v = v_0$ almost everywhere on G_η for any $\eta > 0$. But the function $q = (\bar{\sigma}_{\xi_i}^2 - 2\bar{\sigma}\, \bar{\sigma}_{\xi_i \xi_i})/4\bar{\sigma}^2$ satisfies the conditions of Theorem [17.2] on G_η and our assertion consequently follows from the validity of (17.32).

[18.6]. *All of the weak derivatives of first and second orders of* v_0 *and hence also* u_0 *exist on* G *and are square summable over the compact subsets of* G. *Finally,* u_0 *is an analytic function of its arguments if the* $(\sigma_{x_i}^2 - 2\bar{\sigma}\, \bar{\sigma}_{x_i x_i})/4\bar{\sigma}^2$ *are analytic functions* $(i = 1, \cdots, n)$.

PROOF. According to the preceding theorem it suffices to prove this for a function $v_0 \in D_{\alpha,\varphi}(G)$ satisfying the integral equation

$$
\begin{aligned}
&v_0(x_1, \ldots, x_n) \\
&= \frac{1}{(n-2)s_n} \int \ldots \int_{r \leqslant h} \Delta\left[\frac{1}{r^{n-2}} \psi\left(\frac{r}{h}\right)\right] v_0(\xi_1, \ldots, \xi_n)\, d\xi_1 \ldots d\xi_n \\
&+ \frac{1}{(n-2)s_n} \sum_{i=1}^{n} \int \ldots \int_{r \leqslant h} \frac{1}{r^{n-2}} \psi\left(\frac{r}{h}\right) \frac{\bar\sigma_{\xi_i}^2 - 2\bar\sigma\,\bar\sigma_{\xi_i\xi_i}}{4\bar\sigma^2} v_0(\xi_1, \ldots, \xi_n)\, d\xi_1 \ldots d\xi_n. \qquad (18.18)
\end{aligned}
$$

Inasmuch as the functions

$$
\frac{\bar\sigma_{x_i}^2 - 2\bar\sigma\,\dot{\bar\sigma}_{x_i x_i}}{4\bar\sigma^2} \quad (i = 1, 2, \ldots, n)
$$

are continuously differentiable on any closed bounded domain $G^* \subset \bar G^* \subset G$, our theorem follows from (18.18) and Lemma [17.2.1]. The analyticity of the solution in the indicated case follows from the Remark to Theorem [17.2].

[18.7] *The function* u_0 *satisfies the equation* $L_\alpha(u) = 0$ *almost everywhere on* G.

PROOF. Analogously to [18.4], we have

$$
D_\alpha(u_0, \delta u_0) = 0
$$

for any continuously differentiable variation δu_0 that is identically equal to zero outside a finite domain $G^* \subset G_\eta$, $\eta > 0$. It follows upon integrating by parts that

$$
\int \cdots \int_G L_\alpha(u_0)\, \delta u_0\, dx_1 \ldots dx_n = 0.
$$

Therefore, according to the fundamental lemma of the calculus of variations, $L_\alpha(u_0) = 0$ almost everywhere on G.

We proceed now to an analysis of existence and uniqueness questions for both the differential and variational problems, in the course of which we explain the influence of the exponent α.

§19. The solution of elliptic equations degenerate on the whole boundary of a domain $(0 \leqslant \alpha < 1)$

We will assume from the beginning in this section that G is a finite domain in E^n bounded by an $(n-1)$-dimensional manifold K of smoothness $k \geqslant 2$ (or a finite number of such manifolds) and that there exist constants $\gamma_1 > 0, \gamma_2 > 0$ and $\eta > 0$ such that (see (8.12))

$$\gamma_1 r(P) \leqslant \sigma(P) \leqslant \gamma_2 r(P) \quad \text{for} \quad P \in K^{(n)}_{+\eta}(K). \tag{19.1}$$

As always, $r(P)$ denotes "the distance from a point P to the manifold K along a normal to K." Finally, we will assume that $0 \leqslant \alpha < 1$.

[19.1]. *If* $u \in D_\alpha(G)$ *(see* §18), *then the boundary value*

exists and

$$u|_K \in H_2^{\left(\frac{1-\alpha}{2}\right)}(K)$$

$$\|u\|_{2,K}^{(n-1)} < \infty.$$

This is a special case of the Corollary to Theorem [12.3].

[19.2]. *The following definition of a norm for the functions* $u \in D_\alpha(G)$:

$$\|u\|_{D_\alpha} = \|u\|_{2,K}^{(n-1)} + \sum_{i=1}^{n} \left\| \frac{\partial u}{\partial x_i} \right\|_{2(\sigma,\alpha),G}^{(n)} \tag{19.2}$$

converts the set of admissible functions into a complete normed linear space.

PROOF. The axioms of a norm are verified without difficulty. The finiteness of (19.2) follows from [19.1]. As to completeness, we have from Theorem [12.8] that

$$\|u\|_{2,G}^{(n)} \leqslant c_1 \|u\|_{2,K}^{(n-1)} + c_2 \sum_{i=1}^{n} \left\| \frac{\partial u}{\partial x_i} \right\|_{2(\sigma,\alpha),G}^{(n)} \leqslant c \|u\|_{D_\alpha}. \tag{19.3}$$

It follows from (19.2) and (19.3) that there exist constants $c' > 0$ and $c'' > 0$ such that (see §13, Remark 2°)

$$c' \|u\|_{D_\alpha} \leqslant \|u\|_{W_2^{(1)}(\sigma;\, 0,\alpha)} \leqslant c'' \|u\|_{D_\alpha},$$

i.e. the norms $\|u\|_{D_\alpha}$ and $\|u\|_{W_2^{(1)}(\sigma;0,\alpha)}$ are equivalent. But the space $W_2^{(1)}(\sigma; 0, \alpha)$ is complete. The theorem is proved.

[19.3]. *Suppose a sequence* $u_m \in D_\alpha(G)$ *is a Cauchy sequence in the space* $D_\alpha(G)$, $\lim_{m\to\infty} u_m = u$ *in* $D_\alpha(G)$ *and* $\lim_{m\to\infty} u_m|_K = \varphi$ *in* $L_2^{(n-1)}(K)$ *(the indicated limits exist according to Theorem* [19.2]). *Then* $u|_K = \varphi$.

By virtue of [19.1] and [19.2] this is a special case of Theorem [14.3].

[19.4]. *Suppose given on the boundary* K *of* G *a function* $\varphi \in H_2^{(\rho)}(K)$, $\rho > (1 - \alpha)/2$. *Then the class of* $D_{\alpha,\varphi}$ *admissible functions is nonempty.*

PROOF. This is a special case of Theorem [8.4], according to which the function φ can be extended onto all of E^n as a function $u \in H_2^{(r)}$, where $r = \rho + ½ > 1 - \alpha/2$, such that by virtue of condition 5° of the mentioned theorem (for $s = 1$ and $r = 0$) $D_\alpha(u) < \infty$.

[19.5]. *Suppose the class of* $D_{\alpha,\varphi}$ *admissible functions is nonempty. Then there exists a unique function* $u_0 \in D_{\alpha,\varphi}(G)$ *effecting a minimum of* $D_\alpha(u)$ *in the class* $D_{\alpha,\varphi}(G)$: $D_\alpha(u_0) = d$ *(see* [18.1]).

PROOF. Every minimizing sequence u_m (see [18.2]) is a Cauchy sequence in $D_\alpha(G)$. In fact, by virtue of [19.1] and the Corollary to [18.1] we have

$$\| u_{m+\mu} - u_m \|_{D_\alpha} = \sum_{i=1}^{n} \left\| \frac{\partial u_{m+\mu}}{\partial x_i} - \frac{\partial u_m}{\partial x_i} \right\|_{2\,(\sigma,\alpha)}^{(n)} \to 0$$

$$\text{for } m \to \infty,\ \mu = 1, 2, \ldots .$$

Therefore the completeness of $D_\alpha(G)$ implies that a minimizing sequence u_m has a limit function $u_0 \in D_\alpha(G)$ for which, according to [18.2], $D_\alpha(u_0) = d$.

Suppose there exists another function u such that $D_\alpha(u) - d$. Then from [18.3] we get $D_\alpha(u_0 - u) = 0$, and hence

$$\sum_{i=1}^{n} \left\| \frac{\partial u_0}{\partial x_i} - \frac{\partial u}{\partial x_i} \right\|_{2(\sigma,\alpha)}^{(n)} = 0$$

(see (18.13)); and since, clearly,

$$\| u_0 \underset{K}{-} u \|_2^{(n-1)} = 0,$$

it follows that

$$\| u_0 \underset{G}{-} u \|_2^{(n)} = 0$$

(see (19.3)), which implies $u_0 = u$ almost everywhere on G.

[19.6]. *Suppose the class of* $D_{\alpha,\varphi}$ *admissible functions is nonempty. Then there exists a unique function* $u_0 \in D_{\alpha,\varphi}(G)$ *having weak derivatives up to second order inclusively, belonging to* $L_2(G_\eta)$ *for any* $\eta > 0$ *and such that* $L_\alpha(u_0) = 0$ *almost everywhere on* G.

Following the plan of the proof of Sobolev for the case of Laplace's equation ([2], page 94), we will prove a preliminary lemma. Let

$$P = (x_1, \ldots, x_n),\ Q = (\xi_1, \ldots, \xi_n),\ r = \sqrt{(x_1 - \xi_1)^2 + \ldots + (x_n - \xi_n)^2},\ \eta > 0,$$

$$\omega_h(Q) = \begin{cases} 1 & \text{for } Q \in G_{2h}, \\ 0 & \text{for } Q \notin G_{2h}. \end{cases}$$

and $K(r) = \psi(r/h)/ch^n$ where ψ is defined in §17 (see [17.2]) and $c = s_n \int_0^1 t^n \psi(t)dt$. Then $\int \cdots \int_{r \leq h} K(r) dv_Q = 1$, and $K(r)$ can be regarded as an averaging kernel. We put

$$\chi_h(P) = \int \cdots \int_{r \leqslant h} K(r)\, \omega_h(Q)\, dv_Q.$$

[19.6.1]. *Suppose* $u \in D_\alpha(G)$, $v \in D_\alpha(G)$ *and* $v|_K = 0$. *Then*

$$D_\alpha(u, v) = \lim_{h \to 0} D_\alpha(u, \chi_h v).$$

PROOF. The function χ_h has continuous derivatives of all orders, $\chi_h = 1$ on G_{3h} and $\chi_h = 0$ outside G_h. From the equality

$$\frac{\partial \chi_h}{\partial x_i} = \frac{1}{ch^{n+1}} \int \cdots \int_{r \leqslant h} \psi'\left(\frac{r}{h}\right) \frac{x_i - \xi_i}{r}\, \omega_h(Q)\, dQ$$

it follows that there exists a constant $A > 0$ such that

$$\left|\frac{\partial \chi_h}{\partial x_i}\right| \leqslant \frac{A}{h}, \quad i = 1, 2, \ldots, n.$$

Further,

$$D_\alpha(u, v - \chi_h v) = \int_G \cdots \int \sigma^\alpha \sum_{i=1}^n \frac{\partial u}{\partial x_i} \frac{\partial (v - \chi_h v)}{\partial x_i} dx_1 \ldots dx_n$$

$$= \int_G \cdots \int \sigma^\alpha \sum_{i=1}^n \frac{\partial u}{\partial x_i} \left[\frac{\partial v}{\partial x_i} (1 - \chi_h) - \varphi \frac{\partial \chi_h}{\partial x_i} \right] dx_1 \ldots dx_n$$

$$\leqslant \int_{G \setminus G_{3h}} \cdots \int \sigma^\alpha (1 - \chi_h) \sum_{i=1}^n \frac{\partial u}{\partial x_i} \frac{\partial v}{\partial x_i} dx_1 \ldots dx_n$$

$$+ \int_{G \setminus G_{3h}} \cdots \int \sigma^\alpha v \sum_{i=1}^n \frac{\partial u}{\partial x_i} \frac{\partial \chi_h}{\partial x_i} = I'_h + I''_h.$$

Noting that $|1 - \chi_h| \leqslant 1$ and applying Hölder's inequality, we see that the first integral I'_h does not exceed in modulus

$$\left\{ \int_{G \setminus G_{3h}} \cdots \int \sigma^\alpha \sum_{i=1}^n \left(\frac{\partial u}{\partial x_i} \right)^2 dx_1 \ldots dx_n \right\}^{\frac{1}{2}} \left\{ \int_{G \setminus G_{3h}} \cdots \int \sigma^\alpha \sum_{i=1}^n \left(\frac{\partial v}{\partial x_i} \right)^2 dx_1 \ldots dx_n \right\}^{\frac{1}{2}},$$

which tends to zero as $h \to 0$. For I''_h we have

$$|I''_h| = \left| \int_{G \setminus G_{3h}} \cdots \int \sigma^\alpha v \sum_{i=1}^n \frac{\partial u}{\partial x_i} \frac{\partial \chi_h}{\partial x_i} dx_1 \ldots dx_n \right|$$

$$\leqslant \frac{A}{h} = \int_{G \setminus G_{3h}} \cdots \int \sigma^\alpha |v| \sum_{i=1}^n \left| \frac{\partial u}{\partial x_i} \right| dx_1 \ldots dx_n$$

$$\leqslant \frac{A\sqrt{n}}{h} \left\{ \int_{G \setminus G_{3h}} \cdots \int \sigma^\alpha v^2 dx_1 \ldots dx_n \right\}^{\frac{1}{2}} \left\{ \int_{G \setminus G_{3h}} \cdots \int \sigma^\alpha \sum_{i=1}^n \left(\frac{\partial u}{\partial x_i} \right)^2 dx_1 \ldots dx_n \right\}^{\frac{1}{2}}.$$

Here

$$\lim_{h \to 0} \int_{G \setminus G_{3h}} \cdots \int \sigma^\alpha \sum_{i=1}^n \left(\frac{\partial u}{\partial x_i} \right)^2 dx_1 \ldots dx_n = 0$$

by virtue of the finiteness of $D_\alpha(u)$. On the other hand, the first factor is bounded. To see this we choose an $\eta > 0$ so that for $3h < \eta$ and $P \in G \setminus G_\eta$ there exists a constant $a > 0$ for which $r(P) \leqslant a\rho(P, K)$ (see (8.13)). Then $G \setminus G_{3h} \subseteq K^{(n)}_{3ah} = K^{(n)}_{+3ah}(K)$ (see (8.12)). Applying inequality (19.1) and Theorem [12.5] ($p = 2$; $K = \Gamma$), we get

$$\frac{A\sqrt{n}}{h}\left\{\int_{G\setminus G_{3h}}\!\!\!\cdots\int \sigma v^2 dx_1\ldots dx_n\right\}^{\frac{1}{2}} \leqslant \frac{A\gamma_2^{\frac{\alpha}{2}}\sqrt{n}}{h}\left\{\int_{K_{3ah}^{(n)}}\!\!\!\cdots\int r^{\alpha} v^2 dx_1\ldots dx_n\right\}^{\frac{1}{2}}$$

$$\leqslant 3aAB\gamma_2^{\frac{\alpha}{2}}\sqrt{n}\sum_{i=1}^{n}\left\|\frac{\partial v}{\partial x_i}\right\|_{p(r,\alpha)}^{(n)} \leqslant 3aABn\gamma_2^{\frac{\alpha}{2}} D_{\alpha}^{\frac{1}{2}}(v).$$

Thus $\lim_{h\to 0} I''_h = 0$, and hence $\lim_{h\to 0} D_\alpha(u, v - \chi_h v) = 0$. Q.E.D.

As can be seen in the presence of Theorem [12.5], the cited proof differs from the proof of Sobolev's lemma only by the presence of the weight σ. We have reproduced it here since we will find it convenient in the sequel to refer to the calculation process itself.

We proceed to the proof of Theorem [19.6]. By virtue of Theorem [18.7] it suffices to prove only the uniqueness of a solution of $L_\alpha(u) = 0$ in the class $D_{\alpha,\varphi}(G)$, since the solution u_0 of the variational problem (see [19.5]) has all of the necessary properties. Suppose $u \in D_{\alpha,\varphi}(G)$, the weak derivatives of second order of u exist and are square summable over the compact subsets of G, and $L_\alpha(u) = 0$ almost everywhere on G.

The function $v = u - u_0$ satisfies the conditions of the above lemma. Integrating by parts gives us

$$D_\alpha(u - u_0, \chi_h(u - u_0)) = \int_G\!\cdots\!\int \chi_h(u - u_0)\, L_\alpha(u - u_0)\, dx_1\ldots dx_n = 0.$$

Therefore, according to the lemma, $D_\alpha(u - u_0, u - u_0) = 0$. But then the condition $(u - u_0)|_K = 0$ implies by virtue of inequalities (18.13) and Theorem [12.8] that

$$\| u - u_0 \|_{2}^{(n)} = 0,$$

(with G under the norm) and hence $u = u_0$ almost everywhere on G.

The method of solution of the boundary problem for the case of unbounded domains has its own special peculiarities. We study this problem for the case of the halfspace $G = \overset{+}{E}{}^n = \{(x_i)\colon x_n > 0\}$. Its boundary is the hyperplane $x_n = 0$, which we denote by E^{n-1}. Suppose $\sigma = x_n$, i.e. consider the equation

$$\Delta u + \frac{\alpha}{x_n}\frac{\partial u}{\partial x_n} = 0, \quad 0 \leqslant \alpha < 1, \tag{19.4}$$

which is the Euler equation for the integral

$$D_\alpha(u) = \int \cdots \int_{\overset{+}{E}{}^n} x_n^\alpha \sum_{i=1}^n \left(\frac{\partial u}{\partial x_i}\right)^2 dx_1 \ldots dx_n. \tag{19.5}$$

[19.7]. *If* $u \in D_\alpha(\overset{+}{E}{}^n)$, *then the boundary value* $u|_{E^{n-1}}$ *exists and for any finite domain* $\Gamma \subset E^{n-1}$

$$F|_\Gamma \in H_2^{\left(\frac{1-\alpha}{2}\right)}(\Gamma).$$

This immediately follows from Theorem [11.7] and the Remark to [11.6].

The first peculiarity of the case being considered here consists in the fact that we cannot assert in general that the finiteness of the integral $D_\alpha(u)$ implies the membership of the boundary function in $L_2^{(n-1)}$. Therefore the set $D_\alpha(\overset{+}{E}{}^n)$ in this case cannot be converted into a complete normed linear space.

Consider, however, the set $\widetilde{D}_\alpha(\overset{+}{E}{}^n)$ of all functions $u \in D_\alpha(\overset{+}{E}{}^n)$ such that

$$\| u \|_{\substack{2\\E^{n-1}}}^{(n-1)} < \infty.$$

[19.8]. *The following definition of a norm for the functions* $u \in \widetilde{D}_\alpha(\overset{+}{E}{}^n)$:

$$\| u \|_{\widetilde{D}_\alpha} = \| u \|_{\substack{2\\E^{n-1}}}^{(n-1)} + \sum_{i=1}^n \left\| \frac{\partial u}{\partial x_i} \right\|_{\substack{2(x_n,\,\alpha)\\ \overset{+}{E}{}^n}}^{(n)} \tag{19.6}$$

converts the set $\widetilde{D}_\alpha(\overset{+}{E}{}^n)$ *into a complete normed linear space.*

As in the case of [19.2], the axioms of a norm and the linearity of the space $\widetilde{D}_\alpha(\overset{+}{E}{}^n)$ are verified without difficulty. To prove completeness we note that for any layer $\overset{+}{E}{}^n_a = \{(x_i): 0 < x_n < a\}$

$$\| u \|_{\substack{2\\ \overset{+}{E}{}^n_a}}^{(n)} \leqslant \sqrt{a}\, \| u \|_{E^{n-1}}^{(n-1)} + \frac{1}{\sqrt{1-\alpha}}\, a^{1-\frac{\alpha}{2}} \left\| \frac{\partial u}{\partial x_n} \right\|_{\substack{2(x_n,\,\alpha)\\ \overset{+}{E}{}^n_a}}^{(n)} \tag{19.7}$$

(see [11.2], (11.9) and (11.35); we have somewhat sharpened the constants here). It follows from this inequality that a Cauchy sequence of functions in the sense of the norm (19.6) will be a Cauchy sequence in $W_2^{(1)}(x_n; 0, \alpha; \overset{+}{E}{}^n_a)$ (see §13) and hence will have a limit function $u_\alpha \in W_2^{(1)}(x_n; 0, \alpha; \overset{+}{E}{}_a^n)$ which as $a \to \infty$ tends to a limit function in $\widetilde{D}_\alpha(\overset{+}{E}{}^n)$.

[19.9]. *Suppose given on the hyperplane* E^{n-1} *a function* $\varphi \in H_2^{(\rho)}(\overset{+}{E}{}^{n-1})$, $\rho > (1-\alpha)/2$. *Then the class of* $D_{\alpha,\varphi}$ *admissible functions is nonempty.*

This is a special case of [5.4].

[19.10]. *Suppose the class of* $D_{\alpha,\varphi}$ *admissible functions for the halfspace* $\overset{+}{E}{}^n$ *is nonempty. Then there exists a unique function* $u_0 \in D_{\alpha,\varphi}(\overset{+}{E}{}^n)$ *effecting a minimum of* $D_\alpha(u)$ *in the class* $D_{\alpha,\varphi}(\overset{+}{E}{}^n)$: $\overset{+}{D}_\alpha(u_0) = d$. *The function* u_0 *is an analytic function of its arguments on the halfspace* $\overset{+}{E}{}^n$ *and is a solution of equation* (19.14):

$$\Delta u + \frac{\alpha}{x_n}\frac{\partial u_0}{\partial x_n} = 0.$$

The proof is completely analogous to the proof of Theorems [19.5] and [19.6] with $D_\alpha(G)$ replaced by $\widetilde{D}_\alpha(\overset{+}{E}{}^n)$.

We will not study here the question of the uniqueness of a solution of the boundary problem for the differential equation (19.4) in the case of a halfplane, which was considered in the author's notes [14, 15], since the proof of it is quite tedious. This question is primarily connected with a study of the behavior of the solutions of the equation at infinity (see also in this connection the paper of Huber [41]).

Besides, the notes [14, 15] do not contain the best possible result in this direction. In them the behavior of the solution at infinity is subjected to a restriction which in all probability is implied by the finiteness of the integral $\overset{+}{D}_\alpha(u)$. Apparently, the following theorem holds. *In every nonempty class* $D_{\alpha,\varphi}(\overset{+}{E}{}^n)$ *the equation*

$$\Delta u + \frac{\alpha}{x_n}\frac{\partial u}{\partial x_n} = 0$$

has a unique solution with second derivatives belonging to $L_2(G^*)$ *for any bounded domain* $G^* \subset \overline{G^*} \subset \overset{+}{E}{}^n$. This is easily proved in the case $\alpha = 0$, i.e. for the case of harmonic functions.

§20. The solution of elliptic equations degenerate on part of the boundary of a domain $(0 \leqslant \alpha < 1)$

In this section we consider from the beginning a domain G situated in the halfspace $\overset{+}{E}{}^n = \{(x_i)\colon x_n > 0\}$ and satisfying all of the conditions imposed on this class of domains in §16 (see also §9).

We study the boundary problem for the differential equation

$$\Delta u + \frac{\alpha}{x_n}\frac{\partial u}{\partial x_n} = 0, \tag{20.1}$$

which is the Euler equation for the functional

$$D_\alpha(u) = \int \cdots \int_G x_n^\alpha \sum_{i=1}^{n} \left(\frac{\partial u}{\partial x_i}\right)^2 dx_1 \ldots dx_n. \tag{20.2}$$

[20.1]. *Suppose* $u \in D_\alpha(G)$. *Then the boundary function* $u|_K$ *exists (see* §18), *and for any proper point of the boundary* K *of* G *there exists a neighborhood* U_K *of it such that*

$$u|_{U_K} \in H_2^{\left(\frac{1-\alpha}{2}\right)}$$

In particular, if a proper boundary point belongs to $\overset{+}{K}$ *(see* §9),

$$u|_{U_K} \in H_2^{\left(\frac{1}{2}\right)}.$$

This immediately follows from Theorem [16.1].

As in §19, we could introduce a corresponding normed linear space $D_\alpha(G)$ (see in this connection the Remark to Theorem [16.1]); but for the sake of simplicity of the presentation we confine ourselves to the local point of view, which is sufficient for proving the existence and uniqueness of a solution of the boundary problem.

We recall that the boundary K of G has the form (9.1):

$$K = \bigcup_{\kappa=1}^{\kappa_0} \overline{K}_\kappa \cup \bigcup_{\lambda=1}^{\lambda_0} L_\lambda \cup \bigcup_{\mu=1}^{\mu_0} M_\mu,$$

where the K_κ are manifold pieces, the L_λ are plane manifolds in the hyperplane E^{n-1} and the M_μ are pairwise disjoint closed finite manifolds in the halfspace $\overset{+}{E}{}^n$.

Suppose now a function φ is given on K. Its values on K_κ, L_λ and M_μ will be denoted by $\varphi_{K_\kappa}, \varphi_{L_\lambda}$ and φ_{M_μ} respectively.

[20.2]. *Suppose given on the boundary K of G a function φ such that*

$$\varphi_{K_\kappa} \in \overline{H}_2^{(\rho_\kappa)}(K_\kappa), \quad \rho_\kappa > \frac{n-1}{2}, \quad \kappa = 1, 2, \ldots, \kappa_0,$$

$$\varphi_{L_\lambda} \in \overline{H}_2^{(\rho_\lambda)}(L_\lambda), \quad \rho_\lambda > \frac{n-1}{2}, \quad \lambda = 1, 2, \ldots, \lambda_0,$$

$$\varphi_{M_\mu} \in H_2^{(\rho_\mu)}(M_\mu), \quad \rho_\mu > \frac{1}{2}, \quad \mu = 1, 2, \ldots, \mu_0.$$

Then for any $\alpha > 0$ the class of $D_{\alpha,\varphi}$ admissible functions is nonempty.

This is a rephrasing of Theorem [9.1].

[20.3]. *Suppose the class of $D_{\alpha,\varphi}$ admissible functions is nonempty and $\|\varphi\|_{2\,K}^{(n-1)} < \infty$. Then there exists a unique function $u_0 \in D_{\alpha,\varphi}(\overset{+}{E}{}^n)$ effecting a minimum of $D_\alpha(u)$ in the class $D_{\alpha,\varphi}(\overset{+}{E}{}^n)$: $D_\alpha(u_0) = d$.*

Proof. Suppose u_k is a minimizing sequence, i.e. $\lim_{k\to\infty} D_\alpha(u_k) = d$. Then, as we know (see (18.1)),

$$\left\| \frac{\partial (u_{k+m} - u_k)}{\partial x_i} \right\|_{2(x_n,\alpha)\,G}^{(n)} \to 0 \quad \text{for } k \to \infty, \; m = 0, 1, 2, \ldots, \; i = 1, 2, \ldots, n, \tag{20.3}$$

and since $\| u_{k+m} - u_k \|_{2\,K}^{(n-1)} = 0$, according to [16.2]

$$\| u_{k+m} - u_k \|_{2\,G}^{(n)} \to 0 \quad \text{for } k \to \infty, \quad m = 0, 1, \ldots. \tag{20.4}$$

By virtue of (20.3) and (20.4) the sequence of functions u_k is a Cauchy sequence in the complete space $W_2^{(1)}(x_n; 0, \alpha; G)$ (see (16.4)) and hence has a limit $u_0 = \lim_{k\to\infty} u_k$ in this space. Finally, Theorem [14.3] implies that $u_0|_K = \varphi$; but then from [18.2] we have $D(u_0) = d$.

The uniqueness is proved in the same way as in Theorem [19.5].

[20.4]. *Suppose the class of $D_{\alpha,\varphi}$ admissible functions is nonempty. Then there exists a unique function $u_0 \in D_{\alpha,\varphi}(G)$ which is an analytic function of its arguments and such that*

$$\Delta u_0 + \frac{\alpha}{x_n} \frac{\partial u_0}{\partial x_n} = 0.$$

PROOF. The proof of Theorem [19.6] is completely preserved for the present case with the exception of Lemma [19.6.1], which is changed to account for the presence of corner points and the absence of a degeneracy on the whole boundary but will be completely valid if we prove the boundedness of the integral

$$\frac{1}{h}\left\{\int_{G\setminus G_{3h}}\cdots\int x_n^{\alpha} v^2 dx_1\dots dx_n\right\}^{\frac{1}{2}}$$

for $h \to 0$, where $v \in D_{\alpha}(G), v|_K = 0, G_{3h} = \{P: P \in G, \rho(P, K) > 3h\}$.

Suppose $\eta > 0$ and

$$\mathcal{K}_{\eta}=G\cap\bigcup_{\kappa=1}^{\kappa_0} K_{+\eta}^{(n)}(K_{\kappa}),\quad \mathcal{L}_{\eta}=G\cap\bigcup_{\lambda=1}^{\lambda_0} K_{+\eta}^{(n)}(\overline{L}_{\lambda}),\quad \mathcal{M}_{\eta}=G\cap\bigcup_{\mu=1}^{\mu_0} K_{+\eta}^{(n)}(M_{\mu}). \tag{20.5}$$

Then for each point P belonging to one of the sets in the unions of (20.5) there is uniquely defined the distance $r(P)$ along a normal to a corresponding part of the boundary K. In the sequel, if a point P turns out to belong to several sets in the unions of (20.5), so that several distances $r(P)$ can be considered for it, we will always assume that they all satisfy the appropriate conditions considered below.

We choose an $\eta_0 > 0$ so that when $h < \eta_0/3$ and $P \in G\setminus G_{\eta}$ there exists a constant $a > 0$ for which $r(P) \leqslant a\rho(P, K)$ (see (8.13)). Then

$$G\setminus G_{3h} \subseteq \mathcal{K}_{3ah}\cup\mathcal{L}_{3ah}\cup\mathcal{M}_{3ah}.$$

We further require that η_0 be so small that two different cylindroids in the unions of (20.5) can intersect only if the closures of their bases intersect. We will always assume that $h < \eta_0/3$. Then

$$\frac{1}{h}\left\{\int_{G\setminus G_{3h}}\cdots\int x_n^{\alpha} v^2\, dx_1\dots dx_n\right\}^{\frac{1}{2}} \leqslant \frac{1}{h}\left\{\int_{\mathcal{K}_{3ah}}\cdots\int x_n^{\alpha} v^2\, dx_1\dots dx_n\right\}^{\frac{1}{2}}$$

$$+\frac{1}{h}\left\{\int_{\mathcal{L}_{3ah}}\cdots\int x_n^{\alpha} v^2 dx_1\dots dx_n\right\}^{\frac{1}{2}}+\frac{1}{h}\left\{\int_{\mathcal{M}_{3ah}}\cdots\int x_n^{\alpha} v^2\, dx_1\dots dx_n\right\}^{\frac{1}{2}}.$$

The boundedness of the last integral in the right side of this inequality for $h \to 0$ follows from a lemma of Sobolev ([2], page 94) as well as from [12.5] for $\alpha = 0$, $p = 2$ in view of the fact that its integrand is not degenerate on the corresponding parts M_{μ} of the boundary K $(x_n > x_n^{(0)} > 0)$.

The boundedness of the second integral for $h \to 0$ follows directly from Proposition [11.10].

It remains to estimate an integral of the form

$$\frac{1}{h}\left\{\int_{K_\kappa^{(n)}}\!\!\cdots\!\int x_n^\alpha v^2\,dx_1\ldots dx_n\right\}^{\frac{1}{2}},\quad \kappa = 1, 2, \ldots, \kappa_0,$$

where $K_\kappa^{(n)} = K_{+3ah}^{(n)}(K_\kappa) \cap G$. Suppose $P_0 \in \bar{K}_\kappa \cap E^{n-1}$ and suppose the angle formed by the inward normal N_{P_0} to K_κ at P_0 with the ith coordinate axis is less than a right angle. Then the manifold K_κ admits an explicit representation of the form $x_i = x_i(x_1, \cdots, x_{i-1}, x_{i+1}, \cdots, x_n)$ in a neighborhood of P_0. Therefore by virtue of the smoothness of the manifold K there exists a neighborhood U_{P_0} of P_0 and a number b_{P_0} such that the intersection of a straight line passing through a point of U_{P_0} and parallel to the Ox_i axis with $K_\kappa^{(n)}$ is entirely contained in an interval of length not greater than hb_{P_0}.

Let $K_{\kappa(\eta)} = K_\kappa \cap \{(x_i): 0 < x_n < \eta\}$ and $K_{\kappa\eta}^{(n)} = K_\kappa^{(n)} \cap \{(x_i): 0 < x_n < \eta\}$. From what has been said it follows that there exist an $\eta > 0$, a $b > 0$ and a finite number of pieces $K_{\kappa\nu}$, $\nu = 1, \cdots, \nu_0$, of the manifold K_κ satisfying the following conditions: $K_{\kappa(\eta)} \subset \bigcup_{\nu=1}^{\nu_0} K_{\kappa\nu}$; if $K_{\kappa\nu}^{(n)} = K_{+3ah}^{(n)}(K_{\kappa\nu})$, then $K_{\kappa\nu}^{(n)} \subset \bigcup_{\nu=1}^{\nu_0} K_{\kappa\nu}^{(n)}$, and for each piece $K_{\kappa\nu}$ there exists a number $i = i(\nu)$ such that the intersection with $K_{\kappa\nu}^{(n)}$ of a straight line passing through a point $P \in K_{\kappa\nu}^{(n)}$ and parallel to the Ox_i axis is entirely contained in an interval l_P of length not exceeding bh. To simplify the presentation we will assume essentially (without loss of generality, as is easily seen) that the indicated intersection is an interval l_P. If $i(\nu) < n$, one end of such an interval trivially belongs to K. If $i(\nu) = n$, we take those intervals l_P which do not have an end lying in K and appropriately extend them to K. These newly obtained intervals are again denoted by l_P. It can be assumed without loss of generality that the lengths of the newly obtained l_P also do not exceed bh.

Let $\Delta^{(\nu)} = \bigcup_{P \in K_{\kappa\nu}^{(n)}} l_P \subset K_{\kappa\nu}^{(n)}$. Then

$$K_\kappa^{(n)} \subset \bigcup_{\nu=1}^{\nu_0} \Delta^{(\nu)} \cup \left(K_\kappa^{(n)} \setminus K_{\kappa\eta}^{(n)}\right).$$

The boundedness of the considered integral on the set $K_\kappa^{(n)} \backslash K_{\kappa\nu}^{(n)}$ for $h \to 0$ does not require special considerations, since here $x_n > \eta > 0$, so that we do not have a degeneracy and the necessary estimate follows from the cited lemma of Sobolev

([2], page 94).

Thus the whole question reduces to an investigation of the integrals

$$\frac{1}{h}\left\{\int\limits_{\Delta^{(\nu)}}\!\!\cdots\!\int x_n^{\alpha} v^2\,dx_1\ldots dx_n\right\}^{\frac{1}{2}}.$$

We denote by $\Gamma^{(\nu)}$ the projection of $\Delta^{(\nu)}$ onto the hyperplane $x_i=0$, $i=i(\nu)$, by $l(x_1,\cdots,x_{i-1},x_{i+1},\cdots,x_n)$ the interval of intersection of $\Delta^{(\nu)}$ with the straight line passing through the point $(x_1,\cdots,x_{i-1},0,x_{i+1},\cdots,x_n)\in\Gamma^{(\nu)}$ and parallel to the Ox_i axis and, finally, by

$$\omega_1=\omega_1(x_1,\ldots,x_{i-1},x_{i+1},\ldots,x_n)\text{ and }\omega_2=\omega_2(x_1,\ldots,x_{i-1},x_{i+1},\ldots,x_n)$$

the lower and upper ends respectively of this interval. Clearly, $\omega_1\leqslant\omega_2\leqslant\omega_1+bh$. Since one of the ends of the interval $l(x_1,\cdots,x_{i-1},x_{i+1},\cdots,x_n)$ belongs to K, the relation $v=0$ holds on at least one of them. We will assume for the sake of definiteness that

$$v(x_1,\ldots,x_{i-1},\omega_1,x_{i+1},\ldots,x_n)=0.$$

Suppose first $i(v)<n$, for example $i(v)=1$. Then

$$\frac{1}{h}\left\{\int\limits_{\Delta^{(\nu)}}\!\!\cdots\!\int x_n^{\alpha} v^2\,dx_1\ldots dx_n\right\}^{\frac{1}{2}}$$

$$=\frac{1}{h}\left\{\int\limits_{\Delta(\nu)}\!\!\cdots\!\int x_n^{\alpha}\left|\int\limits_{\omega_1}^{x_1}\frac{\partial v(t,x_2,\ldots,x_n)}{\partial x_1}\,dt\right|^2 dx_1\ldots dx_n\right\}^{\frac{1}{2}}$$

$$\leqslant\frac{1}{h}\left\{\int\limits_{\Delta^{(\nu)}}\!\!\cdots\!\int x_n^{\alpha}\left[\int\limits_{\omega_1}^{x_1}\left(\frac{\partial v(t,x_2,\ldots,x_n)}{\partial x_1}\right)^2 dt\right]\left(\int\limits_{\omega_1}^{x_1}dt\right)dx_1\ldots dx_n\right\}^{\frac{1}{2}}$$

$$\leqslant\frac{\sqrt{b}}{\sqrt{h}}\left\{\int\limits_{\Gamma(\nu)}\!\!\cdots\!\int\int\limits_{\omega_1}^{x_1} x_n^{\alpha}\left(\frac{\partial v(t,x_2,\ldots,x_n)}{\partial x_1}\right)^2\left(\int\limits_{\omega_1}^{\omega_2}dx_1\right)dt\,dx_2\ldots dx_n\right\}^{\frac{1}{2}}$$

$$\leqslant b\left\{\int\limits_{\Delta^{(\nu)}}\!\!\cdots\!\int x_n^{\alpha}\left(\frac{\partial v(t,x_2,\ldots,x_n)}{\partial x_1}\right)^2 dt\,dx_2\ldots dx_n\right\}^{\frac{1}{2}}\leqslant b\,D_\alpha(v). \qquad (20.6)$$

Suppose now $i(v) = n$, $G_\nu = \Delta^{(\nu)} \cap \{(x_i)\colon 0 < x_n < bh\}$, $\Delta_1^{(\nu)}$ is the union of all of the intervals $l(x_1, \cdots, x_{n-1})$ having a nonempty intersection with G_ν, and $\Delta_2^{(\nu)} = \Delta^{(\nu)} \setminus \Delta_1^{(\nu)}$ The necessary estimate of the integral over the domain $\Delta_1^{(\nu)}$ follows at once from [11.10]. For $\Delta_2^{(\nu)}$, denoting by $\Gamma_2^{(\nu)}$ the projection of $\Delta_2^{(\nu)}$ onto the hyperplane $x_n = 0$, we have

$$bh \leqslant \omega_1 \leqslant \omega_2 \leqslant \omega_1 + bh, \quad (x_1, \ldots, x_{n-1}) \in \Gamma_2^{(\nu)},$$

which implies

$$\frac{1}{h}\left\{\int \cdots \int_{\Delta_2^{(\nu)}} x_n^\alpha v^2\, dx_1 \ldots dx_n\right\}^{\frac{1}{2}}$$

$$= \frac{1}{h}\left\{\int \cdots \int_{\Delta_2^{(\nu)}} x_n^\alpha \left|\int_{\omega_1}^{x_n} \frac{\partial v(x_1, \ldots, x_{n-1}, t)}{\partial x_n}\, dt\right|^2 dx_1 \ldots dx_n\right\}^{\frac{1}{2}}$$

$$\leqslant \frac{1}{h}\left\{\int \cdots \int_{\Delta_2^{(\nu)}} x_n^\alpha \left|\int_{\omega_1}^{x_n} \left(\frac{t}{\omega_1}\right)^{\frac{\alpha}{2}} \frac{\partial v(x_1, \ldots, x_{n-1}, t)}{\partial x_n}\right|^2 dx_1 \ldots dx_n\right\}^{\frac{1}{2}}$$

$$\leqslant \frac{1}{h}\left\{\int \cdots \int_{\Delta_2^{(\nu)}} \left(\frac{x_n}{\omega_1}\right)^\alpha \left[\int_{\omega_1}^{\omega_2} t^\alpha \left(\frac{\partial v(x_1, \ldots, x_{n-1}, t)}{\partial x_n}\right)^2 dt\right] (\omega_2 - \omega_1)\, dx_1 \ldots dx_n\right\}^{\frac{1}{2}}$$

$$\leqslant \frac{\sqrt{b}}{\sqrt{h}}\left\{\int \cdots \int_{\Delta_2^{(\nu)}} \left[\int_{\omega_1}^{\omega_2} t^\alpha \left(\frac{\partial v(x_1, \ldots, x_{n-1}, t)}{\partial x_n}\right)^2 dt\right]\left[\int_{\omega_1}^{\omega_2} \left(\frac{x_n}{\omega_1}\right)^\alpha dx_n\right] dx_1 \ldots dx_{n-1}\right\}^{\frac{1}{2}}$$

$$\leqslant \frac{\sqrt{b}}{\sqrt{h}}\left\{\int \cdots \int_{\Gamma_2^{(\nu)}} \left[\int_{\omega_1}^{\omega_2} t^\alpha \left(\frac{\partial v(x_1, \ldots, x_{n-1}, t)}{\partial x_n}\right)^2 dt\right] \left(\frac{\omega_2}{\omega_1}\right)^\alpha (\omega_2 - \omega_1)\, dx_1 \ldots dx_{n-1}\right\}^{\frac{1}{2}}$$

$$\leqslant b\left\{\int \cdots \int_{\Gamma_2^{(\nu)}} \int_{\omega_1}^{\omega_2} t^\alpha \left(\frac{\partial v(x_1, \ldots, x_{n-1}, t)}{\partial x_n}\right)^2 \left(1 + \frac{bh}{\omega_1}\right)^\alpha dx_1 \ldots dx_{n-1}\, dt\right\}^{\frac{1}{2}}$$

$$\leqslant 2^{\frac{\alpha}{2}} b \left\| \frac{\partial v}{\partial x_n} \right\|^{(n)}_{2(x_n, \alpha)\, \Delta_2^{(\nu)}} \leqslant 2^{\frac{\alpha}{2}} b\, D_\alpha(v), \tag{20.7}$$

Q.E.D.

REMARK. As is easily seen, all of the results of this section that were proved for the domain G, with the exception of the sufficient condition for the nonemptiness of the class of $D_{\alpha,\varphi}$ admissible functions, remain valid for the case when the part $\overset{+}{K}$ of the boundary K of G lying in the upper halfspace $x_n > 0$ is piecewise smooth (more precisely, if K is a simple boundary in the terminology of Sobolev ([2], page 72)).

As an example of an infinite domain with a boundary on part of which an elliptic equation has a degeneracy, we consider the layer $\overset{+}{E}{}^n_a = \{(x_i)\colon 0 < x_n < a\}$ whose boundary K consists of the two hyperplanes $E^{n-1} = \{(x_i)\colon x_n = 0\}$ and $E^{n-1}_a = \{(x_i)\colon x_n = a\}$.

[20.5]. *Suppose* $u \in D_\alpha(\overset{+}{E}{}^n_a)$. *Then there exists a boundary function* $u|_K$ *such that for any finite domains* $\Gamma \subset E^{n-1}$ *and* $\Gamma_a \subset E^{n-1}_a$

$$u|_\Gamma \in H_2^{\left(\frac{1-\alpha}{2}\right)}(\Gamma), \quad u|_{\Gamma_a} \in H_2^{\left(\frac{1}{2}\right)}(\Gamma_a).$$

This follows from [12.3].

[20.6]. *Suppose given on the boundary* K *of the layer* $\overset{+}{E}{}^n_a$ *a function* φ *such that*

$$\varphi|_{E^{n-1}} \in H_2^{(\rho)}(E^{n-1}), \rho > \frac{1-\alpha}{2}; \quad \varphi|_{E_a^{n-1}} \in H_2^{(\rho_a)}(E_a^{n-1}), \quad \rho_a > \frac{1}{2}$$

Then the class of $D_{\alpha,\varphi}$ *admissible functions is nonempty.*

This is obtained without difficulty from Theorem [5.4] with the use of [6.1].

[20.7]. *Suppose the class of* $D_{\alpha,\varphi}$ *admissible functions for the layer* $\overset{+}{E}{}^n_a$ *is nonempty. Then there exists a unique function* $u_0 \in D_{\alpha,\varphi}(\overset{+}{E}{}^n_a)$ *effecting a minimum of* $D_\alpha(u)$ *in the class* $D_{\alpha,\varphi}(E^n_a)$.

We note that the convergence of a minimizing sequence is ensured by inequality (19.7). Here it is interesting to observe that the convergence of a minimizing sequence can be ensured by fixing its values on only part of the boundary, in the present case on one of the hyperplanes E^{n-1} or E^{n-1}_a, which follows at once from the indicated inequality. In this case, however, as can be shown by elementary examples, we would not be able to prove the uniqueness of the solution of the corresponding boundary problem for the differential equation.

[20.8]. *Suppose the class of* $D_{\alpha,\varphi}$ *admissible functions is nonempty. Then there exists a unique function* $u_0 \in D_{\alpha\,\varphi}(G)$ *which is an analytic function of its arguments and such that*

$$L_\alpha(u_0) \equiv \sum_{i=1}^{n} \frac{\partial}{\partial x}\left(x_n^\alpha \frac{\partial u_0}{\partial x}\right) = 0. \tag{20.8}$$

Here again it is only necessary to prove the uniqueness of the solution of the boundary problem for equation (20.8). The method by which this is done, although it preserves a well-known analogy with the case of a finite domain, undergoes essential changes connected with the unboundedness of the layer $\overset{+}{E}{}^n_a$.

Thus, suppose $v \in D_\alpha(\overset{+}{E}{}^n_a)$ and

$$v\,|_{E^{n-1}} = v\,|_{E_a^{n-1}} = 0, \tag{20.9}$$

$$L_\alpha(v) = 0. \tag{20.10}$$

Let

$$D_\alpha(u)_E = \int \cdots \int_E x_n^\alpha \sum_{i=1}^{n} \left(\frac{\partial u}{\partial x_i}\right)^2 dx_1 \ldots dx_n$$

and

$$G(h; a^{(1)}, \ldots, a^{(n-1)}; b^{(1)}, \ldots, b^{(n-1)})$$
$$= \{(x_i) : a^{(i)} < x_i < b^{(i)},\ i = 1, 2, \ldots, n-1;\ h < x_n < a - h\}.$$

Suppose given an $\epsilon > 0$. We choose a parallelepiped

$$G_0 = G(3h;\ a_0^{(1)}, \ldots, a_0^{(n-1)};\ b_0^{(1)}, \ldots, b_0^{(n-1)})$$

so that $D_\alpha(v)_{E^n \setminus G_0} < \epsilon$. Further, there exist numbers $a^{(i)} < a_0^{(i)} - 2h$ and $b^{(i)} > b_0^{(i)} + 2h$, $i = 1, \cdots, n-1$, for which

$$\|v\|_{2\ x_i = a^{(i)}}^{(n-1)} < \sqrt{h}, \quad \|v\|_{2\ x_i = b^{(i)}}^{(n-1)} < \sqrt{h}, \quad i = 1, 2, \ldots, n-1, \tag{20.11}$$

where the notation $\|v\|_{2\ x_i = b}^{(n-1)}$ means the norm of v on the set $\{(x_j): x_i = b\} \cap \overset{+}{E}{}^n_a$. This is always true for fixed h by virtue of the condition $v \in L_2^{(n)}(\overset{+}{E}{}^n_a)$ (see (19.7)).

Now let

$$G_{kh} = G(kh,\, a^{(1)} - (k-1)h, \ldots, a^{(n-1)} - (k-1)h;$$
$$b^{(1)} - (k-1)h, \ldots, b^{(n-1)} - (k-1)h),\ k = 1, 2, 3.$$

Then by virtue of the above choice

$$D_\alpha(v)_{E^n_a \setminus G_{3h}}^{+} < \varepsilon. \tag{20.12}$$

As in the proof of Lemma [19.6.1], we consider the functions

$$\omega_h = \begin{cases} 1 & \text{on } G_{2h} \\ 0 & \text{on } E^n \setminus G_{2h} \end{cases}$$

and

$$\chi_h = \frac{1}{ch^n} \int \ldots \int_{r<h} \Psi\left(\frac{r}{h}\right) \omega_h \, dx_1 \ldots dx_n$$

(for the notation see before [19.6.1]). The function χ is infinitely continuously differentiable on E^n, equal to 1 on G_{3h} and equal to zero outside G_h. Therefore by virtue of (20.10)

$$D_\alpha(v, \chi v)_{G_h} = \int \ldots \int_{G_h} L_\alpha(v) \chi v \, dx_1 \ldots dx_n = 0.$$

Hence

$$D_\alpha(v)_{G_{3h}} \leq D_\alpha(v)_{G_h} = D_\alpha(v, v)_{G_h} - D_\alpha(v, \chi v)_{G_h}$$
$$\leq \int \ldots \int_{G_h} x_n^\alpha \sum_{i=1}^{n} \frac{\partial v}{\partial x_i} \left[\frac{\partial v}{\partial x_i}(1-\chi) - v \frac{\partial \chi}{\partial x_i} \right] \Big| \, dx_1 \ldots dx_n$$
$$\leq \int \ldots \int_{G_h \setminus G_{3h}} x_n^\alpha \sum_{i=1}^{n} \left(\frac{\partial v}{\partial x_i}\right)^2 dx_1 \ldots dx_n + \frac{A}{h} \int \ldots \int_{G_h \setminus G_{3h}} x_n^\alpha \sum_{i=1}^{n} |v| \left|\frac{\partial v}{\partial x_i}\right| dx_1 \ldots dx_n. \tag{20.13}$$

According to (20.12) the first integral in the right side of this inequality is less than ϵ. We estimate the second:

$$\frac{A}{h} \int \ldots \int_{G \setminus G_{3h}} x_n^\alpha \sum_{i=1}^{n} |v| \left|\frac{\partial v}{\partial x_i}\right| dx_1 \ldots dx_n \leq \frac{A}{h} \sum_{i=1}^{n} \| v \|^{(n)}_{2(x_n,\, \alpha)\, G_h \setminus G_{3h}} \left\| \frac{\partial v}{\partial x_i} \right\|^{(n)}_{2(x_n,\, \alpha)\, G_h\, G_{3h}}$$
$$\leq \frac{nA}{h} \| v \|^{(n)}_{2(x_n,\, \alpha)\, G_h \setminus G_{3h}} D^{\frac{1}{2}}_\alpha(v) \leq \frac{nA\sqrt{\varepsilon}}{h} \| v \|^{(n)}_{2(x_n,\, \alpha)\, G_h \setminus G_{3h}}. \tag{20.14}$$

We now partition $G_h \backslash G_{3h}$ into parallelepipeds

$$\Delta_1^{(k)} = \{(x_i) : a^{(k)} \leqslant x_k \leqslant a^{(k)} + 2h, \quad a^{(j)} \leqslant x_j \leqslant b^{(j)},$$

$$k \neq j = 1, 2, \ldots, n-1; \quad 3h \leqslant x_n \leqslant a - 3h\},$$

$$\Delta_2^{(k)} = \{(x_i) : b^{(k)} - 2h \leqslant x_k \leqslant b^{(k)}, \quad a^{(j)} \leqslant x_j \leqslant b^{(j)},$$

$$k \neq j = 1, 2, \ldots, n-1; \quad 3h \leqslant x_n \leqslant a - 3h\},$$

$$k = 1, 2, \ldots, n-1.$$

$$\Delta_1^{(n)} = \{(x_i) : a^{(i)} \leqslant x_i \leqslant b^{(i)}, \quad i = 1, 2, \ldots, n-1; \quad h \leqslant x_n \leqslant 3h\},$$

$$\Delta_2^{(n)} = \{(x_i) : a^{(i)} \leqslant x_i \leqslant b^{(i)}, \quad i = 1, 2, \ldots, n-1; \quad a - 3h \leqslant x_n \leqslant a - h\}.$$

Clearly,

$$G_h \setminus G_{3h} \subset \bigcup_{k=1}^{n} \Delta_1^{(k)} \cup \Delta_2^{(k)}.$$

From [11.10] we at once have

$$\| v \|_{2(x_n, \alpha)}^{(n)} {}_{\Delta_\nu^{(n)}} < BD_\alpha(v)\, h, \quad \nu = 1, 2. \tag{20.15}$$

The estimates for the other parallelepipeds are all carried out in the same way. Consider, for example, $\Delta_1^{(1)}$. Using (20.11), we have

$$\| v \|_{2(x_n, \alpha)}^{(n)} {}_{\Delta_1^{(1)}} = \Bigg\{ \int_{a^{(1)}}^{a^{(1)}+h} \int_{a^{(2)}}^{b^{(2)}} \cdots \int_{a^{(n-1)}}^{b^{(n-1)}} \int_{3h}^{a-3h} \Bigg| v(a^{(1)}, x_2, \ldots, x_n)$$

$$+ \int_{a^{(1)}}^{x_1} \frac{\partial v(t, x_2, \ldots, x_n)}{\partial x_1}\, dt \Bigg|^2 dx_1 \ldots dx_n \Bigg\}^{\frac{1}{2}} \leqslant h^{\frac{1}{2}} a^{\frac{\alpha}{2}} \| v \|_{2}^{(n-1)} {}_{x_1 = a^{(1)}}$$

$$+ h \left\| \frac{\partial v}{\partial x_1} \right\|_{2(x_n, \alpha)}^{(n)} {}_{\Delta_1^{(1)}} \leqslant h a^{\frac{\alpha}{2}} + h\varepsilon. \tag{20.16}$$

Substituting (20.15) and (20.16) into (20.14), we get (for $\epsilon < 1$)

$$\frac{A}{h} \int \cdots \int_{G_h \setminus G_{3h}} x_n^{\alpha} \sum_{i=1}^{n} | v | \left| \frac{\partial v}{\partial x_i} \right| dx_1 \ldots dx_n < c\sqrt{\varepsilon},$$

where $c > 0$ is a constant. It therefore follows from (20.13) that

$$D_\alpha(v)_{G_{3h}} < \varepsilon + c\sqrt{\varepsilon}.$$

Combining this with (20.12) gives us

$$D_\alpha(v)_{\overset{+}{E}{}^n_a} = D_\alpha(v)_{G_{3h}} + D_\alpha(v)_{\overset{+}{E}{}^n_a \setminus G_{3h}} < 2\varepsilon + c\sqrt{\varepsilon},$$

which implies, on account of the arbitrariness of $\epsilon > 0$,

$$D_\alpha(v)_{E^n_a} = 0.$$

But then inequality (19.7) implies that $v = 0$ almost everywhere on $\overset{+}{E}{}^n_a$. Q. E. D.

§21. The solution of elliptic equations degenerate on the boundary of a domain with degeneracy exponent $\alpha \geqslant 1$

We again consider the domain G situated in the upper halfspace $\overset{+}{E}{}^n = \{(x_i): x_n > 0\}$ and which was studied in the preceding section. Also, we note that all of the results obtained below are valid, as in §20, for the infinite layer $\overset{+}{E}{}^n_a = \{(x_i): 0 < x_n < a < \infty\}$.

Let $\overset{+}{G}_{(h)} = G \cap \overset{+}{E}{}^n_h$, $\overset{+}{G}_h = G \backslash \overline{G}_{(h)}$ and $\overset{+}{K}_h = K \backslash \overline{G}_{(h)}$. We will assume that the domain $\overset{+}{G}_h$ is regular relative to $\overset{+}{K}_h$ for sufficiently small h (see [12.7]).

Then the following inequality holds by virtue of [16.2] for any function u defined on G together with its weak derivatives of first order:

$$\| u \|^{(n)}_{2,\overset{+}{G}_h} \leqslant a_1 \| u \|^{(n-1)}_{2,\overset{+}{K}_h} + a_2 h^{-\frac{\alpha}{2}} \sum_{i=1}^{n} \left\| \frac{\partial u}{\partial x_i} \right\|^{(n)}_{2(x_n,\alpha),G}, \tag{21.1}$$

where the constants a_1 and a_2 depend only on the shape of the domain G.

We again consider the functional

$$D_\alpha(u) = \int \cdots \int x_n^\alpha \sum_{i=1}^{n} \left(\frac{\partial u}{\partial x_i} \right)^2 dx_1 \ldots dx_n$$

and its Euler equation

$$L_\alpha(u) \equiv \sum_{i=1}^{n} \frac{\partial}{\partial x_i}\left(x_n^\alpha \frac{\partial u}{\partial x_i}\right) = 0,$$

but under the condition $\alpha \geqslant 1$.

As we will see below, the definition of the class of admissible functions undergoes a modification in this case.

[21.1]. *Suppose $D_\alpha(u) < \infty$. Then there exists a boundary function $u|_{\overset{+}{K}}$ belonging to the class $H_2^{1/2}$ in a neighborhood $U_{\overset{+}{K}}(P)$ of every point $P \in \overset{+}{K}$.*

This is obtained from Theorem [12.3], for example, when $p = 2$, $\lambda = 0$ and $\alpha = 0$, i.e. in the absence of degeneracy.

We note that, in general, the function u cannot be assigned values in the sense of convergence in the mean on the part of the boundary K of G lying in the hyperplane $x_n = 0$.

Suppose now a function φ is given on $\overset{+}{K}$. A function u defined on G is said to be $D_{\alpha,\varphi}$ admissible if $u|_{\overset{+}{K}} = \varphi$ and $D_\alpha(u) < \infty$. The class of all $D_{\alpha,\varphi}$ admissible functions is denoted by $D_{\alpha,\varphi}(G)$, while the class of all $D_{\alpha-\epsilon,\varphi}$ admissible functions is denoted by $D_{\alpha-\epsilon,\varphi}(G)$.

[21.2]. *Suppose given on the boundary $\overset{+}{K} = \bigcup_1^{\kappa_0} K_\kappa \cup \bigcup_1^{\mu_0} M_\mu$ (see (9.1)) a function φ such that*

$$\varphi_{K_\kappa} \in \overline{H}_2^{(\rho_\kappa)}(K_\kappa), \quad \rho_\kappa > \frac{n-1}{2}, \quad \kappa = 1, 2, \ldots, \kappa_0;$$

$$\varphi_{M_\mu} \in H_2^{(\rho_\mu)}(M_\mu), \quad \rho_\mu > \frac{1}{2}, \quad \mu = 1, 2, \ldots, \mu_0.$$

Then the class $D_{\alpha,\varphi}(G)$ is nonempty for any $\alpha \geqslant 1$.

This follows from Theorem [9.1].

[21.3]. *Suppose the class of $D_{\alpha,\varphi}$ admissible functions is nonempty. Then there exists a unique function $u_0 \in D_{\alpha,\varphi}(G)$ effecting a minimum of $D_\alpha(u)$ in the class $D_{\alpha,\varphi}(G)$.*

PROOF. We note that by virtue of (21.1) every minimizing sequence is a Cauchy sequence in the complete space $W_2^{(1)}(x_n; 0, \alpha; \overset{+}{G}_h)$ and hence has a limit u_0 in this space for any $h > 0$. It readily follows that $u_0 \in D_{\alpha,\varphi}(G)$ and $D_\alpha(u_0) = d$.

The uniqueness of u_0 is also obtained at once on the basis of inequality (21.1) and Theorem [18.3].

[21.4]. *Suppose the class of $D_{\alpha,\varphi}$ admissible functions is nonempty. Then there exists a function $u_0 \in D_{\alpha,\varphi}(G)$ which is an analytic function of its arguments and satisfies the equation*

$$\sum_{i=1}^{n} \frac{\partial}{\partial x_i}\left(x_n^{\alpha} \frac{\partial u}{\partial x_i} \right) = 0$$

Moreover, any class $D_{\alpha-\epsilon,\varphi}(G)$, $\epsilon > 0$, contains at most one such function u_0.

PROOF. The uniqueness of u_0 in a class $D_{\alpha-\epsilon,\varphi}(G)$ is proved in the same way as the uniqueness in Theorem [20.4]. It is only necessary to replace the references to the inequality of [11.10] by a reference to [11.13], in which α and $\alpha + \epsilon$ are replaced by $\alpha - \epsilon$ and α respectively when $p = 2$. We note further that estimates (20.6) and (20.7) hold for any $\alpha > 0$.

Finally, the Corollary to [11.12] implies that for fixed $x_1, \cdots, x_{n-1}$

$$u_0(x_1, \ldots, x_n) = o\left(\frac{1}{x_n^{\frac{\alpha}{2} - \frac{1}{2} + \varepsilon}} \right) \quad \text{as } x_n \to 0,$$

provided $u_0 \in D_{\alpha-\epsilon,\varphi}(G)$, $\epsilon > 0$.

BIBLIOGRAPHY

1. M. V. Keldyš and M. A. Lavrent'ev, *On the stability of solutions of the Dirichlet problem*, Izv. Akad. Nauk SSSR Ser. Mat. **1** (1937), 551–593. (Russian; French summary)

2.* S. L. Sobolev, *Applications of functional analysis in mathematical physics*, Izdat. Leningrad. Gos. Univ., Leningrad, 1950; English transl., Transl. Math. Monographs, vol. 7, Amer. Math. Soc., Providence, R. I., 1963. MR **14**, 565; 29 #2624.

3.* S. M. Nikol'skiĭ, *Inequalities for entire functions of exponential type and their application to the theory of differentiable functions of several variables*, Trudy Mat. Inst. Steklov. **38** (1951), 244–278; English transl., Amer. Math. Soc. Transl. (2) **80** (1969), 1–38. MR **14**, 32.

4.* ———, *Properties of certain classes of functions of several variables on differentiable manifolds*, Mat. Sb. **33** (75) (1953), 261–326; English transl., Amer. Math. Soc. Transl. (2) **80** (1969), 39–118. MR **16**, 453.

5. ———, *On Dirichlet's problem for the circle and half-space*, Mat. Sb. **35** (77) (1954), 247–266. (Russian) MR **16**, 589.

6. ———, *On the solution of the polyharmonic equation by a variational method*, Dokl. Akad. Nauk SSSR **88** (1953), 409–411. (Russian) MR **15**, 425.

7. V. I. Kondrašov, *On the theory of nonlinear and linear problems on characteristic values*, Dokl. Akad. Nauk SSSR **90** (1953), 129–132. (Russian) MR **15**, 41.

8. ———, *The behavior of functions from* L_p^ν *on manifolds of different dimensions*, Dokl. Akad. Nauk SSSR **72** (1950), 1009–1012. (Russian) MR **14**, 149.

9. ———, *On boundary problems with degenerate boundary for some nonlinear operator equations*, Candidate's Dissertation, Moscow State University, 1942. (Russian)

10. ———, *On the theory of boundary problems and eigenvalue problems in domains with degenerate boundary for variational and differential equations*, Doctoral Dissertation, Steklov Inst. Math., Moscow, 1948. (Russian)

11. L. M. Slobodeckiĭ and V. M. Babič, *On boundedness of the Dirichlet integral*, Dokl. Akad. Nauk SSSR 106 (1956), 604–606. (Russian) MR 17, 959.

12. R. Courant and D. Hilbert, *Methoden der mathematischen Physik*. Vol. I, Springer-Verlag, Berlin, 1931; English transl., Interscience, New York, 1943, 1953; Russian transl., GITTL, Moscow, 1951. MR **5**, 97; 13, 800; 16, 426.

---

* *Translator's note.* All page references to [2], [3] and [4] are to the English translations.

13. L. D. Kudrjavcev, *On extension of functions and imbedding of classes of functions*, Dokl. Akad. Nauk SSSR **107** (1956), 501–504. (Russian) MR **17**, 1190.

14. ———, *On the solution by the variational method of elliptic equations which degenerate on the boundary of the region*, Dokl. Akad. Nauk SSSR **108** (1956), 16–19. (Russian) MR **19**, 283.

15. ———, *On a generalization of S. M. Nikol'skiĭ's theorem on the compactness of classes of differentiable functions*, Uspehi Mat. Nauk **9** (1954), no. 1 (59), 111–120. (Russian) MR **16**, 453.

16. ———, *On an integral inequality*, Naučn. Dokl. Vysš. Školy Fiz.-Mat. Nauki **1959**, no. 3, 25–32. (Russian) RŽMat **1961** #6E 29.

17. Ja. S. Bugrov, *On imbedding theorems*, Dokl. Akad. Nauk SSSR **116** (1957), 531–534. (Russian) MR **20** #4101.

18. F. Tricomi, *Sulle equazioni lineari alle derivate parziali di 2° ordine, di tipo misto*, Atti Accad. Naz. Lincei Rend. C1. Sci. Fis. Mat. Natur. **14** (1923), 133–247; Russian transl., OGIZ, Moscow, 1947.

19. M. A. Lavrent ev and A. V. Bicadze, *On the problem of equations of mixed type*, Dokl. Akad. Nauk SSSR **70** (1950), 373–376. (Russian) MR **11**, 724.

20. A. V. Bicadze, *On the problem of equations of mixed type*, Trudy Mat. Inst. Steklov. **41** (1953). (Russian) MR **16**, 43.

21. M. V. Keldyš, *On certain cases of degeneration of equations of elliptic type on the boundary of a domain*, Dokl. Akad. Nauk SSSR **77** (1951), 181–183. (Russian) MR **13**, 41.

22. I. N. Vekua, *On a generalization of the Poisson integral for a half-plane*, Dokl. Akad. Nauk SSSR **56** (1947), 229–231. (Russian) MR **9**, 187.

23. M. I. Višik, *Boundary-value problems for elliptic equations degenerating on the boundary of a region*, Mat. Sb. **35** (77) (1954), 513–568; English transl., Amer. Math. Soc. Transl. (2) **35** (1964), 15–78. MR **16**, 927.

24. S. G. Mihlin, *On the theory of degenerate elliptic equations*, Dokl. Akad. Nauk SSSR **94** (1954), 183–185. (Russian) MR **16**, 367.

25. A. Huber, *On the uniqueness of generalized axially symmetric potentials*, Ann. of Math. (2) **60** (1954), 351–358. MR **16**, 258.

26. P. Germain and R. Bader, *Problemes elliptiques et hyperboliques pour une equation du type mixte*, O.N.E.R.A. Publ. No. 60 (1952).

27. B. A. Fuks, *Theory of analytic functions of several complex variables*, Fizmatgiz, Moscow, 1948; English transl. of rev. ed. in 2 parts, Transl. Math. Monographs, vols. 8, 14, Amer. Math. Soc., Providence. R. I., 1963, 1965. MR 12, 328; **29** #6049; **32** #5915.

28. N. I. Ahiezer, *Lectures on the theory of approximation*, OGIZ, Moscow, 1947; English transl., Ungar, New York, 1956. MR **10**, 33; **20** #1872.

29. G. M. Fihtengol'c, *A course in differential and integral calculus*. Vol. 1, Fizmatgiz, Moscow, 1958; German transl., VEB Deutscher Verlag, Berlin, 1968. MR **39** #1a.

30. H. Whitney, *Analytic extensions of differentiable functions defined in closed sets*, Trans. Amer. Math. Soc. 36 (1934), 63–89.

———, *Differentiable functions defined in arbitrary subsets of Euclidean space*, Trans. Amer. Math. Soc. **40** (1936), 309–317.

31. M. R. Hestenes, *Extension of the range of a differentiable function*, Duke Math. J. **8** (1941), 183–192. MR **2**, 219.

32. V. M. Babič, *On the extension of functions*, Uspehi Mat. Nauk 8 (1953), no. 2 (54), 111–113. (Russian) MR **15**, 110.

33. L. D. Kudrjavcev, *On the homology groups of topological spaces*, Candidate's Dissertation, Moscow State University, Moscow, 1948. (Russian)

34. S. M. Nikol'skiĭ, *Boundary properties of functions in regions with angles*, Dokl. Akad. Nauk SSSR 111 (1956), 26–28. (Russian) MR **18**, 795.

35. ———, *Boundary properties of functions defined on a region with angular points*. I, Mat. Sb. **40** (**80**) (1956), 303–318; English transl., Amer. Math. Soc. Transl. (2) 83 (1969), 101–120. MR **18**, 723.

———, *Boundary properties of functions defined on a region with angular points*. II: *Harmonic functions on rectangular regions*, Mat. Sb. **43** (**85**) (1957), 127–144; English transl., Amer. Math. Soc. Transl. (2) 83 (1969), 121–141. MR **20** #5368a.

———, *Boundary properties of functions defined on a region with angular points*. III: *Connection with the polyharmonic problem*, Mat. Sb. **45** (87) (1958), 181–194; English transl., Amer. Math. Soc. Transl. (2) 83 (1969), 143–157. MR **20** #5368b.

36. N. M. Gunther, *La theorie du potentiel et ses applications aux problemes fondamentaux de la physique mathematique*, Gauthier-Villars, Paris, 1934; Russian transl., GITTL, Moscow, 1953; English transl., Ungar, New York, 1967. MR **16**, 357.

37. S. N. Bernšteĭn, *On best approximation of continuous functions by polynomials of given degree*, Soobšč. Har'kov. Mat. Obšč. (2) 13 (1912), 47–194; rev. reprint, in *Collected works*. Vol. 1: *The constructive theory of functions* (1905–1930), Izdat. Akad. Nauk SSSR, Moscow, 1952, pp. 11–104; abridged French transl., Acad. Roy. Belg. Cl. Sci. Mém. (2) **4** (1912), fasc. 1. MR **14**, 2.

38. G. H. Hardy, J. E. Littlewood and G. Pólya, *Inequalities*, Cambridge Univ. Press, New York, 1934; Russian transl., IL, Moscow, 1948. MR **18**, 722.

39. S. N. Bernstein, *Sur la nature analytique des solutions des equations aux derivees partielles du second ordre*, Math. Ann. **59** (1904), 20–76.

40. I. G. Petrovskiĭ, *Sur l'analyticité des solutions des systemes d'equations differentielles*, Mat. Sb. **5** (47) (1939), 3–70. MR **1**, 236.

41. A. Huber, *A theorem of Phragmen-Lindelöf type*, Proc. Amer. Math. Soc. **4** (1953), 852–857. MR **15**, 877.

42. M. I. Višik, *The method of orthogonal and direct decomposition in the theory of elliptic differential equations*, Mat. Sb. **25** (**67**) (1949), 189–234. (Russian) MR **11**, 520.

43. L. D. Kudrjavcev, *Extensions of functions and embedding of function classes. Application to the solution by variational methods of elliptic equations degenerating on the boundary*, Doctoral Dissertation, Steklov Inst. Math., Moscow, 1956. (Russian)

44. S. M. Nikol'skiĭ, *Extension of functions of several variables preserving differential properties*, Mat. Sb. **40** (82) (1956), 243–268; English transl., Amer. Math. Soc. Transl. (2) **83** (1969), 159–188. MR **19**, 398.

45. A. P. Calderon and A. Zygmund, *On the existence of certain singular integrals*, Acta Math. **88** (1952), 85–139. MR **14**, 637.

46. N. Aronszajn, *Boundary values of functions with finite Dirichlet integrals*, Conference on Partial Differential Equations, University of Kansas, Lawrence, Kan., 1955.

47. E. Gagliardo, *Caratterizzazioni delle tracce sulla frontiera relative ad alcune classi di funzioni in n variabili*, Rend. Sem. Mat. Univ. Padova **27** (1957), 284–305. MR **21** #1525.

48. L. N. Slobodeckiĭ, *Estimates of solutions of elliptic and parabolic systems*, Dokl. Akad. Nauk SSSR **120** (1958), 468–471. (Russian) MR **21** #5060.

49. A. A. Vašarin, *The boundary properties of functions having a finite Dirichlet integral with a weight*, Dokl. Akad. Nauk SSSR **117** (1957), 742–744. (Russian) MR **20** #1113.

50. P. I. Lizorkin, *Boundary properties of a certain class of functions*, Dokl. Akad. Nauk SSSR **126** (1959), 703–706. (Russian) MR **21** #5700.